Business Strategy and Corporate Governance in the Chinese Consumer Electronics Sector

CHANDOS
ASIAN STUDIES SERIES:
CONTEMPORARY ISSUES AND TRENDS

Series Editor: Professor Chris Rowley,
Centre for Research on Asian Management, Cass Business School,
City University, UK; HEAD Foundation, Singapore
(e-mail: c.rowley@city.ac.uk)

Chandos Publishing is pleased to publish this major Series of books entitled *Asian Studies: Contemporary Issues and Trends*. The Series Editor is Professor Chris Rowley, Director, Centre for Research on Asian Management, City University, UK and Director, Research and Publications, HEAD Foundation, Singapore.

Asia has clearly undergone some major transformations in recent years and books in the Series examine this transformation from a number of perspectives: economic, management, social, political and cultural. We seek authors from a broad range of areas and disciplinary interests: covering, for example, business/management, political science, social science, history, sociology, gender studies, ethnography, economics and international relations, etc.

Importantly, the Series examines both current developments and possible future trends. The Series is aimed at an international market of academics and professionals working in the area. The books have been specially commissioned from leading authors. The objective is to provide the reader with an authoritative view of current thinking.

New authors: we would be delighted to hear from you if you have an idea for a book. We are interested in both shorter, practically orientated publications (45,000+ words) and longer, theoretical monographs (75,000–100,000 words). Our books can be single, joint or multi-author volumes. If you have an idea for a book, please contact the publishers or Professor Chris Rowley, the Series Editor.

Dr Glyn Jones
Chandos Publishing
E-mail: gjones@chandospublishing.com
www.chandospublishing.com

Professor Chris Rowley
Cass Business School, City University
E-mail: c.rowley@city.ac.uk
www.cass.city.ac.uk/faculty/c.rowley

Chandos Publishing: Chandos Publishing is an imprint of Woodhead Publishing Limited. The aim of Chandos Publishing is to publish books of the highest possible standard: books that are both intellectually stimulating and innovative.

We are delighted and proud to count our authors from such well known international organisations as the Asian Institute of Technology, Tsinghua University, Kookmin University, Kobe University, Kyoto Sangyo University, London School of Economics, University of Oxford, Michigan State University, Getty Research Library, University of Texas at Austin, University of South Australia, University of Newcastle, Australia, University of Melbourne, ILO, Max-Planck Institute, Duke University and the leading law firm Clifford Chance.

A key feature of Chandos Publishing's activities is the service it offers its authors and customers. Chandos Publishing recognises that its authors are at the core of its publishing ethos, and authors are treated in a friendly, efficient and timely manner. Chandos Publishing's books are marketed on an international basis, via its range of overseas agents and representatives.

Professor Chris Rowley: Dr Rowley, BA, MA (Warwick), DPhil (Nuffield College, Oxford) is Subject Group leader and the inaugural Professor of Human Resource Management at Cass Business School, City University, London, UK, and Director of Research and Publications for the HEAD Foundation, Singapore. He is the founding Director of the multi-disciplinary and internationally networked Centre for Research on Asian Management (http://www.cass. city.ac.uk/cram/index.html) and Editor of the leading journal Asia Pacific Business Review (www.tandf.co.uk/journal/titles/13602381.asp). He is well known and highly regarded in the area, with visiting appointments at leading Asian universities and top journal Editorial Boards in the UK, Asia and the US. He has given a range of talks and lectures to universities, companies and organisations internationally with research and consultancy experience with unions, business and government, and his previous employment includes varied work in both the public and private sectors. Professor Rowley researches in a range of areas, including international and comparative human resource management and Asia Pacific management and business. He has been awarded grants from the British Academy, an ESRC AIM International Study Fellowship and gained a 5-year RCUK Fellowship in Asian Business and Management. He acts as a reviewer for many funding bodies, as well as for numerous journals and publishers. Professor Rowley publishes extensively, including in leading US and UK journals, with over 370 articles, books, chapters and other contributions.

Bulk orders: some organizations buy a number of copies of our books. If you are interested in doing this, we would be pleased to discuss a discount. Please e-mail wp@woodheadpublishing.com or telephone +44 (0) 1223 499140.

Business Strategy and Corporate Governance in the Chinese Consumer Electronics Sector

HAILAN YANG AND STEPHEN L. MORGAN

CP
CHANDOS
PUBLISHING

Oxford Cambridge Philadelphia New Delhi

Chandos Publishing
Hexagon House
Avenue 4
Station Lane
Witney
Oxford OX28 4BN
UK
Tel: +44 (0) 1993 848726
E-mail: info@chandospublishing.com
www.chandospublishing.com

Chandos Publishing is an imprint of Woodhead Publishing Limited

Woodhead Publishing Limited
80 High Street
Sawston
Cambridge CB22 3HJ
UK
Tel: +44 (0) 1223 499140
Fax: +44 (0) 1223 832819
www.woodheadpublishing.com

First published in 2011

ISBN:
978 1 84334 656 2
(Chandos Publishing)

ISBN:
978 0 85709 152 9
(Woodhead Publishing)

Typeset by RefineCatch Limited, Bungay, Suffolk

Contents

List of figures

List of tables

Abbreviations and acronyms

AMC	Asset management company
BoD	Board of directors
BS	Business strategy
CCP	Chinese Communist Party
CE	Consumer electronics
CG	Corporate governance
CMRS	Contract management responsibility system
COE	Collectively-owned enterprise
CSRC	China Securities Regulatory Commission
CTV	Colour TV
DPOE	Domestic privately-owned enterprise
FDI	Foreign direct investment
GDP	Gross domestic product
IC	Integrated circuit
IPO	Initial Public Offering
IT	Institutional theory
JV	Joint venture
LP	Legal person
M&A	Merger and acquisition
MNC	Multi-national corporation
NPL	Nonperforming loan
NSB	National Statistics Bureau
OEM	Original Equipment Manufacturing
R&D	Research and development
RBV	Resource-based view
SAIC	State Administration for Industry and Commerce of China

SASAC	State-owned Assets Supervision and Administration Commission
SB	Supervisory board
SOE	State-owned enterprise
SSE	Shanghai Stock Exchange
SZSE	Shenzhen Stock Exchange
TCE	Transaction cost economics
TVE	Township and village enterprise
WTO	World Trade Organization

Preface

Three decades of economic reform has allowed China to emerge on the global stage in a way that has been dramatic and astonishing for many observers. From a country largely irrelevant to global trade and investment, China became the second largest economy in 2010. As a result of the reform, economic globalization and increasingly fierce market competition, Chinese firms now face a more complex and volatile institutional environment at home. Many have also embarked on adventures abroad – outward foreign investment from China has soared during the past decade. Focusing on Chinese firms at home, in this book we ask: What are the respective roles of new rules of the market economy and what is the remaining legacy of the planned economy on Chinese firms? How do the Chinese firms improve their competitive advantage internally to meet the challenges of reform? These questions have recently fascinated researchers of organization and strategy. However, what makes the questions in China particularly complex is the very character and pace of change. Continuously changing market conditions in transition economies challenge the theoretical and methodological approaches to studying business strategies (BS) and corporate governance (CG) that have been derived from advanced and (relatively) stable market economies.

The transition economies therefore provide a new context in which to consider the relative strengths and weaknesses of the different theoretical concepts. As the largest and one of the fastest growing transition economies in the world, China provides a significant counter-example to the findings of existing literature on law, finance and growth that examine strategy and governance of firms. In many ways, it is difficult to compare the corporate reform in China with that in developed countries, or even in Eastern Europe and the Soviet Union for that matter. Unlike some other communist countries, where the government undertook massive centralized privatization of the State-owned enterprises, one striking feature of the Chinese transition is its fast growth without large-scale direct privatization. This has often been put down to Chinese pragmatism, summarized in the phrase 'crossing the river by feeling for the rocks' (*mozhe shitou guohe*).

Although the Chinese State still maintains ultimate control of the economy, there has been a deeper movement away from State planning and State-run institutions. Chinese reform has succeeded in creating more competitive markets, thereby providing a fairer market environment for all types of firms than in the pre-reform period. In the transition from a centrally planned to a market-oriented economy, China has experienced the emergence and prosperity of entrepreneurial activities over the past three decades. Chinese manufacturers have had more freedom and willingness to innovate according to market liberalization, competitive pressures and consumer demand. Increased incentives have enabled managers to be more market- and learning oriented and to enhance the competitive position of their firms. Business strategy and corporate governance are two of the major internal means to achieve the competitive advantage of firms. Chinese managers are recognizing and reacting to competitive pressures. The increasing autonomy of managerial decision-making in strategies represents a genuine transformation of the business environment.

Business strategy is a necessary but not a sufficient condition for a competitive approach. The governance of the firm also matters for competitiveness. The form of corporate governance Chinese listed firms have embraced is unique. We can see elements of both the Anglo-American and the German–Japanese models of corporate governance, yet the bulk of the listed firms in China are State-controlled and the State is the largest shareholder in the typical listed firm. With the adoption of an increasingly market-oriented system in China, market competition has driven firms to dilute the ownership control of the State in exchange for public funds to finance their growth, which has resulted in an ownership evolution, in part. However, the reform related to corporate governance is relatively harder and more sensitive in China because it is about who controls and owns the property of the firms, which for most listed firms was originally a State asset. Transformation of firms in China has more readily begun with changing their business strategy, which had little to do directly with their ownership. The reform of business strategy in China preceded the drive to reform corporate governance. Contrary to some of the literature on the topic, the reform or change in the role of corporate governance is not necessarily the most direct way to improve firm performance.

The interrelationship between the institutional environment, business strategies, competitive position and corporate governance is the root of the different processes that influence firm development in China during the transition period. We explain the differences in the process of the development of firms in terms of a virtuous circle of positive feedback in response to market stimulus or a vicious circle of defensive and negative feedback that reinforced anti-market orientations associated with their administrative heritage before the advent of economic reforms. During the process, the reform of business strategy may drive the reform of corporate governance. China has a diverse, complex and rapidly changing economy. The dynamic evolution of the business strategy and corporate governance in China will remain a fascinating research agenda well into the future.

Many people deserve to be thanked for their assistance in preparing this book. A very special word of thanks to the University of Melbourne for giving us support that has enabled us to undertake this study. We are grateful for the participating companies, the Hisense, Panda, Huadong, Haier, Chunlan, Yankon, Tiantong and Silan groups, for generously providing us with access to senior staff, information and research material. We would like to express our heartfelt gratitude to many colleagues and friends, who graciously gave their time and knowledge to provide us with ideas, comments and suggestions. Without their patient assistance, the completion of this book would not have been possible. Foremost, we are grateful to our families. Their support and love gave us the strength and inspiration to write this book.

We would like to thank the following for permission to reproduce copyright material: Figure 1.1, Chapter 1, reprinted with the permission of Edward Elgar Publishing, from *The Development of Corporate Governance*, by On Kit Tam. Table 3.1, Chapter 3, reprinted with the permission of Nova Science Publishers, from *China's Industries in Transition*, by Xiaojuan Jiang. Figure 5.2, Chapter 5, reprinted with the permission of Blackwell Publishing, from *Exit the Dragon? Privatization and State Control in China*, by Guy Liu and Pei Sun.

Introduction to strategy, corporate governance and corporate reform

The fundamental problem of China research is that the concepts and theories favoured in the disciplines were developed out of assumptions about systems which operate quite differently from China, which is more a civilization pretending to be a nation-State.

(Lucian Pye)

Abstract: As the largest and fastest growing transition economy in the world, China's economic achievement has received much attention. Economic development over the past three decades has been astonishing. From a country largely irrelevant to global trade and investment, China became the second largest economy in 2010. Among the many developments has been the transformation of enterprise organization and ownership. Our focus in this book is to first analyse how Chinese firms have developed their business strategies and corporate governance; and second, the interrelationship between the two at this point in the economic reform process.

The aim of this chapter is to provide a framework for the book. It begins with an overview of Chinese economic reform and specifically the reform of Chinese firms. The Chinese consumer electronics sector is used for the study, partly on the basis of its significance in national economy, but also because its growth experience is typical of many industries in the Chinese economy as a whole. Next, the business strategy (BS) and corporate governance (CG) of Chinese firms is described and research issues identified. During the transition process, managers have progressively come to exercise their greater decision-making power and to introduce a modern corporate system in order to enhance performance of the firm. Finally we outline the research objectives and the methods used, and conclude with a summary of the structure of the book.

Key words: strategy, governance, reform, China.

Overview

The Chinese transition from a planned economy to a market-oriented economy over the past three decades has seen considerable research focused on the reform of firms.[1] Research has examined questions such as: how do the Chinese firms improve their competitive advantage using internal resources to meet the challenges of the reforms? What is the respective role of new rules of the market economy and remaining legacy of the planned economy on Chinese firms? And how have managers sought to develop organizational structures to compete more effectively in the emerging market? This book aims to analyse how Chinese firms in the consumer electronics (CE) sector have developed their business strategy and corporate governance – two aspects of their internal organization crucial for competitive advantage – during the reform process. Chinese transition to an increasingly market-oriented economy has given birth to a new diversity in firm ownership types. The book focuses on three broad ownership forms: the State-owned, collectively-owned and domestic privately-owned firms. As will become apparent, there is frequently difficulty in clearly classifying firm ownership structure in China. These three firm types best enable a comparative analysis of business strategy and corporate governance in this industry sector, which has been one of the leading sectors in the growth of Chinese-produced consumer manufactured products for domestic and export markets (OECD, 2002; CCIDConsulting, 2007).

Development of business strategy and the appropriate form of corporate governance are two of the major internal means to achieve the competitive advantage of the firms (Child and Pleister, 2003; Filatotchev and Toms, 2003). Competitive advantage comes from either lowering costs relative to competitors (doing the same better or more efficiently) or differentiation (do something different that provides more value to the user than the competitor), which enables the firm to outperform others in the sector (Porter, 1985). The transition from a central planned to a market-oriented economy in China has increased the autonomy of managerial decision-making in strategies that represents a genuine transformation of the business (Tan, 2002; Tan and Tan, 2003). Chinese firms in general have become more market responsive and in turn competitive. The strategic decision-making capability of managers should further be analysed with reference to the firm's governance arrangements (Dalton et al., 1999; Filatotchev and Toms, 2003; Thomsen and Pedersen, 2000), which are related to the abilities of managers, their incentives to develop effective strategies and their accountability to stakeholders. Driven by the fierce market competition in the sector, firms have strived to reform their corporate governance (Liu and Woo, 2001). Improved corporate governance, and greater rewards for managers, are seen as the means to better enable the managers first to exploit strategically the internal resources of the firm, and second to position the firm to better explore external resources, thus to improve the competitive advantage of the firm (Jefferson and Su, 2006; Thomsen and Pedersen, 2000).

Past empirical studies of Chinese firms have largely neglected the link between the three elements of business strategy, corporate governance, and performance in a transition economy such as that of China. These elements are related in a series of cause–effect interactions that start with the transition of the business environment and the response of firms in allocating resources. Changes in these elements may cause a fit or misfit between the strategic choices of firms and the external environment (Grant, 1999). Analysing the interaction between various internal and external elements helps enrich our understanding of the different processes that influence firm growth in the CE sector in China during the transition period.

We examine the CE sector in particular because it is one of the country's most important and dynamic manufacturing sectors; it was also one of the earliest market-oriented sectors. Thirty years ago the sector did not exist. China is now one of the world's biggest producers and exporters of CE products (CCID Consulting Report, 2008). The sector is therefore useful for examining how firms develop their business strategy and corporate governance in China, which is undergoing the transition to a market-oriented economy domestically in which firms face challenges from multi-national corporations (MNCs) globally. Many Chinese firms in this sector over the past decade have also been pioneers in venturing abroad, setting up subsidiaries in both advanced market economies and developing economies.

For Chinese managers who have been so enthusiastic to embrace Western theories and practices for the past 30 years, the challenge is how to adopt and implement approaches to management in a very different institutional and political context to that in which they were developed.

This book investigates the generalizability of the Western model of management in China. The transfer, adaptation and implementation of Western management practice is not only important for Chinese firms, which are eager to integrate themselves with the world, but also important for MNCs, which have tremendous incentives to succeed in the Chinese domestic market as local consumers acquire increasing levels of discretionary income.

Business strategy and the corporate governance of Chinese firms

Empirical studies on business strategies

Empirical studies of business strategies in transition economies are relatively sparse. Modern literature on strategy in the second half of the twentieth century was dominated by a focus on Western firms mostly in advanced market economies (Child and Pleister, 2003; Peng, 2005; Tsui et al., 2004). The role of emerging economies' operations as part of a 'global strategy' is occasionally discussed, but the literature typically treats the subject from the perspective of multinationals pondering major corporate incursions into or retreats from foreign environments (Nelson, 1990). Strategy within the business environment of emerging countries has been largely ignored. Yet the challenges in these environments are huge. Emerging economies are in a state of dynamic flux, whereas business strategies based on Western industrialized economies assume that the business and institutional environment is sufficiently predictable to make medium and long-term plans reliable.

The heterogeneity and dynamic changes found in emerging economies magnify the challenge to the wholesale adoption of developed economy-based theoretical and methodological approaches in these countries (Wright et al., 2005). For example, it is difficult to compare the enterprise reform in China with European countries because of the different approach to the respective market-institutional environment and because of the very different historical, economic and cultural contexts (Buck et al., 2000). Thus, the theories driving the strategy research agenda are not equally effective in or even relevant for studying different economies (Wright et al., 2005).

Until recently, China was not on the radar screen of research of business strategy (Tsui, 2004; Peng, 2005). The few management journals that considered the problems of management research in China called for a new approach.[2] The dynamics of the Chinese market raises questions about how firms use resources and the performance consequences in changing environment (Zhou and Li, 2007). In the Chinese context, the environment–strategy relationship holds, as it did in a study of Tan and Litschert (1994) in the same setting more than a decade earlier (Tan and Tan, 2004). However, the optimal choice of firm strategy had changed in terms of the deployment of resources and the choice of developing alternative resources and capabilities (Tan and Tan, 2004; White and Xie, 2006). More recent developments of business strategies in China emphasize the 'strategic flexibility' of firms, which depends jointly on the inherent flexibility of resources available to the organization, and on the flexibility of managers in applying those resources to alternative courses of action (Wright et al., 2005).

Empirical studies on corporate governance

Although the role of corporate governance has been a central issue in the management literature of developed economies for some time, many findings from developed countries should not be considered conclusive or universally applicable if the transition economies have not been taken into account (Hovey, 2004). Patterns of corporate governance and control differ significantly across countries because of national differences, including those related to the structures of the ownership and the composition of corporate boards (Jenkinson and Mayer, 1992; Roe, 1990). For instance, the major issues of corporate governance in developed countries concern legal rules on how to limit the agency problem, protect shareholders and creditors, and provide room for managerial initiative. However, a special concern in China is the role of the State in the governance of the firm because it often remains the largest shareholder even after quasi-privatization via a public listing on a stock exchange (Qian, 2002).

After 30 years of gradual reform in China, corporate governance has become not only the centre of the enterprise reform but a focus of the capital market as well (Bai et al., 2004; Sun et al., 2002). It has been at the top of the agenda for Chinese enterprise reform since the early 1990s (Tam, 1999; Tang and Ward, 2003).

The ownership structure that has evolved in many Chinese listed companies in their reasonably short history is unique. Despite making a public issue of stock, overall the governments at central, provincial and local levels largely determine the ownership structure of firms. At the time of listing, the State typically held a significant proportion of shares (Qu, 2003). Besides direct control through ownership of a dominant holding in many firms, the State extends its ownership using indirect controls. The government uses the control chain including State solely-owned companies, State-controlled non-listed companies, State-controlled publicly listed companies and State-owned academic institutions to control listed firms (Liu and Sun, 2003). All of these ostensibly State-related parties hold shares that are not traded and cannot be acquired by private investors. This suggests that the main feature of the structure of large and publicly-listed Chinese firms is that of State dominance. Being the largest shareholder in a company does not necessarily mean absolute control of the firm if there exist sufficient large stakes held by the other large shareholders (Liu and Sun, 2003). But the situation is less likely in Chinese listed firms, in which the largest shareholder always owns a sufficiently large number of shares.

China's Company Law does not mention any disclosure obligation on the part of directors or any specific liabilities for directors who fail to perform their obligations ('Chinese Company Law', 2005). In theory, the primary functions of the board of directors are to monitor managerial functions, assist management by providing advice, and veto any management decisions that are deemed harmful to shareholders (Kose and Senbet, 1998; Monks and Minnow, 2001). In China, however, the board of directors has fewer powers compared with those in advanced economies. The main task of the board of directors is to implement resolutions passed at the shareholders' meeting (Chen, 2005). Directors are mainly nominated by major shareholders and elected to the board by the shareholders' general meeting (Clark, 2003; Denis, 2001; Estrin, 2002). In listed State-owned enterprises (SOEs), one of the directors is typically the Party secretary, who represents the Party committee (Bebchuk and Roe, 2004; Cheng and Lawton, 2005; Morch et al., 2005). The close relationship between the government and listed SOEs continues. A company's performance and directors' personal interests and rewards are not closely related. The directors can easily harm the company's interests in pursuit of their own (Cheng and Lawton, 2005).

The corporate governance in China has elements of the German two-tier codetermination model with a board of directors and a small supervisory board. Considering the fact that the government is the major shareholder of many listed companies, the two-tier board system in China clearly shows the government's intervention in day-to-day business operations (Chen, 2005). The Chinese practice is somewhat different from that found in Germany. In Germany, supervisory board members include representatives of large institutional shareholders (such as banks that provide share capital as well as loans to companies), along with union representatives. The supervisory board in China consists largely of communist party representatives and party-sanctioned union officials, whose task is primarily to protect State interests, not those of other shareholders or community shareholders (Mallin and Rong, 1998; Tam, 1999). The dependence of supervisors on the executive board and Chinese Communist Party (CCP) has led to their inability to actually supervise directors and managers (Tenev et al., 2002). Supervisory boards rarely contest decisions made by boards of directors and company executives. Therefore, the supervisory function has no real powers in China, and supervisory boards typically have little input and influence. However, there are two opposing points of view related to the function of a supervisory board in China. One holds that the supervisory board is ineffective and a single-tier board structure should be adopted (Tenev et al., 2002). The other argues that its potential as a monitoring mechanism should be improved (Dahya et al., 2003; Tam, 1999). Corporate governance and hence performance will inevitably improve if the supervisory board is improved (Dahya et al., 2003).

Most Chinese public companies have a corporate governance structure made up of many of the characteristics of a public corporation in mature market economies. Yet, these borrowed institutions and structures are often in reality embellishments that do not necessarily perform the same role as they do in the models that they emulate (Hovey, 2005).

Figure 1.1 shows clearly the various points at which the government or the Party has input into Chinese corporate governance system. In effect, the Party-State dominates all levels of corporate governance and management of a firm.

Different paths of development of firms

The development of business strategy and the appropriate ownership structure are major internal means to achieve the competitive advantage of firms (Child and Pleister, 2003; Filatotchev and Toms, 2003). According to debates on the growth of Chinese firms, two schools of argument have come into being: the ownership school and the management school (Qu, 2003). The 'ownership school' argues that the form of ownership – and especially its reform – is the key to the development of the firms. State ownership is held to be intrinsically less effective than private ownership,

Figure 1.1 Structures of China's corporate governance system

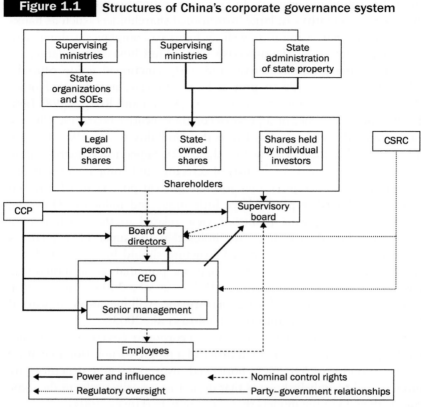

Source: Tam (1999: 100).

mostly because politicians force State-controlled firms to pursue political goals or other social objectives rather than profit maximization. Therefore the key to the reform is to diversify State ownership, in order to eliminate government control of the firms. Not so the view of the 'management school', which emphasizes the need to improve the management of firms, such as the development of a market-oriented strategy without the shackles of State demands (Liu and Garino, 2001b). This school of thought does not believe there is anything intrinsically inefficient about State ownership itself. Ownership of the firm is an irrelevant concept if a firm is regarded as a Williamsonian nexus of contracts (Fama, 1980). State-controlled firms are not different from those listed firms in market economies that have widespread public ownership, and a firm's performance depends on its management culture and the clarity of its goals and objectives (Chang and Singh, 1997; Wortzel and

Wortzel, 1989). Therefore, according to the management school, the solution for the inefficient State-controlled firms is to grant the firms more managerial autonomy and allow them to adopt more commercially oriented business strategies (Liu and Garino, 2001b), which would positively influence future performance.

Objectives of the book

Since the early 1980s, the Chinese government has tried to restructure the incentive system for firms, increase autonomy for managers, reform products and factor markets, and reform the ownership structure of State firms (Liu and Woo, 2001). These reforms were intended to change firms, especially State-owned enterprises (SOEs), into more commercially-oriented businesses (Zhang and Parker, 2004). Such reforms were imposed externally by the State. As economic reform in China advances, the practices of firms in the marketplace have provided important research topics in business strategy and public policy. In this sense, how firms reform themselves internally is an important issue. Business strategy and the pattern of corporate governance are the two major internal reform measures needed to improve the performance of firms.

The overall purpose of this book is to examine business strategy and corporate governance and their relationship for Chinese firms in the CE sector at this juncture in the reform process. The specific aims of the study are as follows:

1. Review the business strategies of Chinese firms in the CE sector.
2. Examine the practices of Chinese firms in the area of corporate governance in the CE sector.
3. Identify different paths of development of firms in the Chinese context.
4. Enable policy issues to be better informed.

The following section discusses these objectives in further detail.

Objective 1: Exploring the business strategies of Chinese firms in the consumer electronics sector

Business strategy comprises the actions of managers to achieve short- and long-term goals based on the competitive resources available to firms (Grant, 1999). The transition from central planning to a market-oriented economy has increased the autonomy of managerial decision-making strategies, which represents a genuine transformation of business (Tan and Litschert, 1994). China has recently been included in the strategy research. This is because the country has emerged from being a peripheral member of the global economy to that of a core contributor (Peng, 2005). Although Chinese business research has much to offer to global strategy research, the work done so far is still thin.

One of the purposes of the book is to study the evolution of learning strategies by which competitive firms in the Chinese CE sector have adapted to the dynamic changes of the institutional environment. Over the years since their founding, these firms have been able to broaden and deepen their set of resources and capabilities to respond to the changing environment. Thus, the cases try to provide insights into the interactions among key factors of environmental changes, exploitative and explorative strategies, and performance. In addition, the book also tries to explain why the strategies of some firms do not fit with their environments, which leads to their poorer competitive position.

Objective 2: Exploring the practices of Chinese firms in the area of corporate governance

A strategic decision should be analysed with reference to the firms' governance arrangements (Filatotchev and Toms, 2003), which are related to the abilities of managers, their incentives to develop effective strategies and their accountability to stakeholders. Managers' actions can be aligned to shareholders' interests by bonding managers contractually, monitoring them, and/or providing them with appropriate incentives (Denis, 2001). Fierce market competition in the CE sector is driving reform of corporate governance in these firms (Liu and Woo, 2001). Improved corporate governance is seen as the means to better enable Chinese managers to exploit strategically the internal resources of the firm and position the firm to better tap external resources.

Objective 3: Exploring two different paths of development of firms in China

Business strategy and corporate governance are important determinants of the competitive advantage of firms in market economies. The Chinese CE sector has experienced dramatic changes related to these two issues.

Owing to the different competitive position of the firms, the study attempts to combine research on the most important elements of business strategy and corporate governance in order to develop a model describing different processes of firms' development during Chinese transition. The model aims to outline the causal relationship among various elements that enhance or detract from the competitive position of the firms.

Objective 4: Proposing policy recommendations on alternatives for China

The final objective of the book is to formulate insights that might assist the development of policy recommendations. Although this study concerns only one industry, the CE sector epitomizes the dynamism of the Chinese industry today. Other industrial sectors in China have undergone a similar transformation as China has moved from a planned to a market-oriented economy. The Chinese government is pressing firms to improve business strategy and corporate governance. The findings in this book for the CE sector can be extended to other manufacturing sectors in China. The empirical and theoretical findings from this study will provide recommendations for policy-making on the reform of firms in China.

To summarize, business strategy and corporate governance and their social ramifications are significant concerns for current reform policies in China. However, there are insufficient empirical studies. One of the purposes of the book will be to provide some policy recommendations on alternatives for China.

Research methodology

The case study method

As it is an explorative study, semi-structured interview questions are used to explore comprehensively the reality of the firms studied, with a view of developing theoretical hypotheses based on a priori assumptions derived from earlier literature and the field observation (Ghauri and Gronhaug, 2005; Yin, 1984). This book aims to find empirical evidence in order to support or reject the findings from previous research. The project foci are the business strategies and governance structure of firms. For this reason, this study adopts a qualitative methodology that is shaped by its objective. Qualitative researching uses a variety of methods including case studies, ethnographies, narratives, discourse analyses, and symbolic interactions studies (Huberman and Miles, 2002). The case study is a research strategy that focuses on understanding the dynamics present within a single setting, such as a firm (Eisenhardt, 1989). A case study is therefore appropriate for this study. The case study is but one of a range of qualitative research methods, which we explore in more detail below.

The case study method is accepted as a scientific tool in management research (Gummesson, 2000; Yin, 2003). For the purposes of this book, the case study is adopted as the research method for the following reasons.

First, in this book, variable measurements related to resources, transaction costs and institutional factors pose problems that present difficulties for strategy research in transition economies (Hoskisson et al., 2000). For example, measures for resources and transaction costs are difficult to obtain from secondary data or to test empirically (Rindfleisch and Heide, 1997). Similarly, measures of institutional factors are difficult to construct, especially in a transition economy whose environment is changing rapidly. These difficulties related to measuring variables in the research imply that a case-based method is a better way to investigate and explore strategies and corporate governance knowledge in a transition economy.

Second, a case study method has been chosen because the research stresses the circumstantial context that exemplifies the institutional environment under which the firms have developed. A case study contributes uniquely to the understanding of a complex social phenomenon (Yin, 1984, 2003). According to Yin (2003), a case study is used when the phenomenon is not easily distinguishable from its context. The situation of Chinese firms occurs in a very complex social context. Since qualitative research is an inquiry process of understanding that explores a social problem, this book can use the cases to explore the complexity of the social context which the firms are facing in the transitional stage, and thereby help the researcher to understand how the various parts of an event are linked and interconnected within the social setting. Therefore, this method can meet the requirements of the research topic.

Third, statistics may provide broad indicators that have a limited power to explain the complicated internal workings of firms (Gummesson, 2000). Owing to the complexity of ascertaining the actual roles of various elements during the development of Chinese firms, it is likely that the study of the relationship between various variables on the basis of aggregate statistics is sometimes descriptive, making it hard to decide whether there is a causal relationship between them (Jiang, 2001c). If researchers want an in-depth understanding of the mechanism of change, they need not study a large number of cases. Thus the case study method is an effective way to understand the experiences, administrative environment and industrial administration of the sample of the firms in this study. In particular, the cases enable comparative studies of SOEs, collectively-owned enterprises (COEs) and domestic privately-owned enterprises (DPOEs), including longitudinal and historical data crucial to their development.

Fourth, a case study approach offers the researcher the opportunity to ask the respondents their opinions about the events and their insights into managerial practices. The views of the interviewees on business strategy and corporate governance of their firms may be vital to understanding the growth process of a firm. Direct contact with the creators or actors of firms (managers and workers) on the ground is the best way to learn about them. Through talking with people, asking questions and listening, and analysing their use of language and construction of discourse, enables the researcher not only to explore the outcome of the change through which the firms have passed, but also to understand participants' insights into the detailed workings of the relationships between firm practices and social processes. According to this perspective, meanings and understandings are created in an interaction, which is effectively a co-production, involving researcher and interviewees (Mason, 2002).

Fifth, the cases in this study fall into different categories of ownership. The diversification of ownership cannot be taken for granted. The ownership structure varies even within the same category of companies in China. There is a danger, especially in the Asian context, of form taking precedence over content (Backman, 1999). For example, in the case of corporate governance reforms, it is not uncommon for Chinese firms to adopt some mechanisms – for example, the supervisory board – without knowing its purpose, and thus without the expected result (Hovey, 2004). These internal workings of firms related to business strategies and corporate governance are better grasped using a case-by-case analysis.

Measures of the competitive position

Two variables of the competitive position of firms are examined in this study – market share and the growth rate of annual sales. In developed nations, accounting profit and stock returns are two of the major indicators of a company's financial performance. However, new strategies and competitive realities demand new measurement systems (Eccles, 1991). There is a shift from treating financial measures as the foundation for performance measures, to placing non-financial measures such as quality and market share on an equal footing with financial performance measures (Stainer and Heap, 1996). Other authors (Johnson and Kaplan, 1987; Kaplan and Norton, 1992) lay out arguments against judging performance based solely on financial criteria.

Jiang (2001c) points out that not all the major changes in Chinese economic development can be accounted for from the perspective of the economic system. For instance, falling profitability may result from the emergence of competition from the non-State enterprises, which is a desirable effect of economic reform (Jefferson and Rawski, 1994; Naughton, 1995). Competition has caused government-monopolized profits to drop. A drop in the profitability of the manufacturing industry is an inevitable stage in reform. But this does not mean that reform has failed to improve efficiency (Jiang, 2001b). Moreover, managers have both the incentive and discretion to manipulate the financial account. For example, the profit figures have been inflated or fabricated by the management to support the stock price (Feinerman, 2007). Firms have also shown a loss in order to obtain a subsidy from the government (Sicular, 1995). In order to avoid misleading conclusions based on potentially unreliable financial data in the Chinese context, this study assesses the performance of firms in the form of their competitive position as measured by their market share and the growth rate of sales of the companies.

Market share refers to the percentage of the overall volume of business in a given market that is controlled by one company in relation to its competitors. The important factor in computing the relative market share is not the exact number associated with the sales volume. The position relative to the competition is more important (Cook, 1995). It is easier to measure market share than some other common objectives, such as maximizing profits. Ambler and Wang (2002) compare the performance measures employed in China and the UK, and find that more Chinese respondents than UK respondents considered market share as an important marketing metric. The focus on market share results from the transition to the era of hyper-competition in China. Facing the increasingly intensified competition, market share became more concentrated during 1990s (Schlevogt, 2000). Firms improve their production efficiency in a drive to increase the market share of their products (Brandt and Zhu, 2005).

However, a company may be tempted to set too low a price in order to achieve a higher market share. To remedy the weakness of the market share, a second competitive position indicator chosen is the growth rate of annual sales. Information on a firm's relative market share should be correlated with the growth rate of the firms. If both the company's competitive position and growth rate are strong, then the company is in a fortunate position and is known as a 'star' (Urban and Star, 1991).

Data sources

There were two steps in the collection of information for this study. First, the secondary sources of information such as company reports and published information in both the Chinese media and Western media were reviewed. Based on these data, a list of questions was developed, focusing on strategies, corporate governance and their interrelationship. Second, a series of semi-structured interviews were conducted with senior managers and board members in companies, and government officials and professionals in order to achieve triangulation. The interviews sought to obtain opinions about factors that influenced a firm's business strategy and corporate governance.

Review of documentary materials

The secondary sources related to the case studies included public information and internal company information. There were two principal sources of public documents: a) annual reports, circulars and other company documents; b) public reporting about the company from academic journals, newspapers, business magazines, yearbooks, almanacs and websites. These data were used to develop a company profile and to trace the company's recent history and deals.

In-depth interviews

One-on-one interviews with a variety of informants were conducted in China. For each company, we interviewed senior managers and members of the board. The selection of informants was aimed at collecting data from those who were in a good position to be informed about the firm's business strategies and corporate governance.

The approach to the interviews

Interviews were conducted in Mandarin and, on average, took around two hours. The interviews were conducted as follows:

1. A series of 56 semi-structured interviews was conducted. The interviews were face-to-face, carried out both in an informal style as well as a formal question-and-answer format. The researcher interviewed senior managers working in different parts of the enterprise including the Chief Technology Officer, independent board members, the secretary of the board, the director of the Party department of the firms, the director of human resources and administration management, and the director responsible for marketing. In addition, local government organizations and consulting firms as well as local supervisory agencies, commercial units, finance bureaus, planning departments and other relevant entities were also interviewed. This process provided a check against distortions based on any single interview.

    Some informants were interviewed more than once. Interviews were tape-recorded unless informants objected. The first round of interviews was conducted during five months between March and August 2006. An additional 28 interviews were conducted between April and July 2007.

2. The interviews discussed specific situations, decisions and actions, not just the general opinions of interviewees. The interview notes for each enterprise constituted the main data for the project. Detailed notes were taken at all interviews. The notes were subdivided according to interviewees. Although notes were not wholly transcribed verbatim, the record generally followed the flow of conversation at each interview so as to preserve the character of the notes as primary data.

The interviewees

The profile of the interviewees is outlined in Tables 1.1 and 1.2 (more detail on each individual is introduced in Part IV of the Bibliography). The interviewees of each company in Table 1.1 were senior executives, board members including executive directors, non-executive directors, independent directors and supervisory board members. A total of 56 persons from the companies were interviewed.

The adoption of different types of business strategies and corporate governance of the target firms was conducted in different phases in this study. Therefore the questions relating to the earlier period were normally

Table 1.1   Profile of interviewees from companies

Company	Interviewee role	Number	Average years in current position	Average years with company
Hisense	Senior company executives	3	10	14
	Board member, non-executive	2	8	19
	Board member, independent	1	6	N/A
	Supervisory board member	1	5	11
Panda	Senior company executives	3	8	20
	Board member, non-executive	2	11	18
	Board member, independent	1	7	N/A
	Supervisory board member	1	4	8
Huadong	Senior company executives	3	5	23
	Board member, executive	1	4	26
	Board member, non-executive	1	4	25
	Board member, independent	1	3	N/A
	Supervisory board member	1	5	30
	Board secretary	1	4	18
Haier	Senior company executives	1	6	17
	Board member, executive	2	4	23
	Supervisory board member	1	4	N/A
	Board secretary	1	4	9
Chunlan	Senior company executives	2	5	22
	Board member, executive	1	5	24
	Board member, non-executive	1	5	28
	Board member, independent	1	4	N/A
	Supervisory board member	1	5	17
	Board secretary	1	5	20
Yankon	Senior company executives	3	5	11
	Board member, executive	3	4	9
	Supervisory board member	1	5	10
	Board secretary	1	3	6
Silan	Senior company executives	2	6	8

(*continued*)

Table 1.1	Profile of interviewees from companies (*cont'd*)

Company	Interviewee role	Number	Average years in current position	Average years with company
	Board member, executive	3	11	10
	Supervisory board member	1	5	8
	Board secretary	1	3	12
Tiantong	Senior company executives	2	3	8
	Board member, executive	2	3	7
	Board member, independent	1	6	N/A
	Supervisory board member	1	7	9
	Board secretary	1	3	10

Note: In the column 'Average years with the company' N/A means that the person is not affiliated with or employed by the company.

Table 1.2	Profile of non-company interviewees

Department	Interviewee role	Number	Years with company	Years on current position
Policy and Regulation Office, Shandong SASAC	Deputy director	1	15	3
Shandong Securities Co.	Senior consultant	1	14	6
Qingdao Pale Consulting Company	Senior consultant	1	20	8
Shandong Electronics Bureau	Vice president	1	34	10
Huarong Asset Management Co., Beijing	Assistant to general manager	1	18	5
Shandong Securities Co.	Senior manager	1	12	5

answered by the interviewees who had worked in the company for a long time, covering the different stages of the development of the firm. In addition to the company informants, six non-company persons were also interviewed (see Table 1.2). The six people were from the local government, an asset management company, a securities company, a consulting company and the local Electronics Bureau. The number of interviewees varied with the size of the company and the availability of participants.

According to Company Law in China, a board of directors should consist of between five and 19 directors. The average size of boards was 9.9 (Tenev et al., 2002). Our case study firms averaged 9.1 directors. It has to be noted that during the interviews, the composition of the board of directors in Chinese companies had to be taken into account. Two-thirds of all directors were executive directors. About half of the executive directors had senior management positions. Interviews with the board of directors included both executive directors and independent directors.

Six board secretaries were interviewed. According to Chinese Company Law (1994) and Chinese Securities Law (1998), the board of directors (BoD) secretary, the equivalent of the company secretary in Western countries, is one of the senior officers, and reports to the BoD. The 1994 'Mandatory Provision for Articles of Association of Companies to be Listed Overseas' in China (included in Chinese Company Law, 1994) introduced the BoD secretary as a new position within listed companies. In the management structure of Chinese listed firms, the position of BoD secretary is similar in standing to the position of managing director, and the individual is expected to be knowledgeable about a listed firm (Nee and Opper, 2006). The BoD secretary is the firm's spokesman and coordinates the meetings of shareholders, board of directors, supervisory board and managers. The BoD secretary is also involved in making management decisions and providing directors with legal assistance about the implementation of the resolutions of the board. Quite a few studies have used the BoD secretary as a respondent to answer questions related to the decision-making power of shareholders, managers, board of directors and State actors (Chang and Wong, 2003; Nee and Opper, 2006). Among the six BoD secretaries interviewed for this study, two of them (Interviews 15, 33) were also board members.

Structure of the book

The book is divided into seven chapters. The first chapter provides an overview of the framework of the book. Chapter 2 reviews the relevant research literature. Its objective is to analyse the body of theory and empirical research in terms of its relevance for the present study. It also identifies the weaknesses that this book aims to address. This chapter briefly discusses overall theoretical and philosophical perspectives as well as fully developing the research theory.

Chapter 3 discusses the background of Chinese economic development. This chapter examines the evolution of the enterprise reforms in China. In doing so it highlights the complexity of the context of this book and provides a broad understanding of the background and reforms in China. The chapter puts forward a four-stage scheme of development in the CE sector. Next, the chapter analyses the development of enterprises with different ownership and the shareholding system in China. The book draws on this review in formulating the methodology, conclusions and policy advice.

Chapters 4 and 5 analyse the business strategy and corporate governance of Chinese firms respectively. Chapter 4 uses the results of the case study to elucidate the coevolution of strategies with changes in the industry and the overall Chinese business environment. Discussion of this part is based on March's (1991) categories of exploitation versus exploration strategies. Managers must allocate resources to different types of activities: those that exploit a firm's existing resources and capabilities, and those that explore new resources and capabilities. This chapter integrates institutional theory, the resource-based view and transaction cost economics to analyse exploitation versus exploration strategies.

Chapter 5 compares the corporate governance of the case study firms. Ownership structure and characteristics of the boards are its focus. In the course of implementing reforms to adapt to the market, the reforms have changed the corporate landscape of all types of firms. However, this chapter finds that government involvement in firms will still exist for some time in the Chinese context, which is characterized by the transition period. This chapter, together with the previous chapter (Chapter 4) allows the reader to gain an understanding of the development and reform of Chinese firms in these two areas. These two chapters also provide a basis for the discussion in the following chapter about the interrelationship between the two areas.

Chapter 6 attempts to draw out the model of institutional factors, business strategies, corporate governance and performance under the Chinese transition context, summarizes the characteristics of firms with different ownership, and predicts the trend of the development of Chinese firms. The contributions and limitations of the study are also outlined in this chapter.

Chapter 7 concludes the book, positioning the findings of the study within the context of existing research. Some practical recommendations are offered to Western and Chinese managers interested in improving their competitiveness in China. The bibliography lists Chinese sources of reference, English sources of reference and detailed information about the interviewees.

Notes

1. In this book, 'Chinese firms' or 'China' refers to firms in the People's Republic of China or Mainland China.
2. The major motivation for establishing the journal *Management and Organization Review* in 2004 was the need to provide an outlet for management research that considered the internal dynamics of the discourse on management in China, rather than simply seeing China as a site where Western management was experienced.

Business strategy and corporate governance: theoretical and empirical perspectives

To engage pro-active restructuring strategies, managers need incentives and strategic advice. As a result, strategy content is influenced by the firm's ownership structure and board characteristics. In a dynamic perspective, corporate governance factors may be affected by strategic choices.

(Igor Filatotchev and Steve Toms)

Abstract: The development of business strategy and appropriate forms of corporate governance are two of the major means to achieve competitive advantage. Depending on the type of institutional environment a firm faces, strategy can either drive or constrain performance. Business strategy is a necessary but not sufficient condition for competitive performance. Performance is also influenced by a firm's governance arrangements, which are related to the abilities of the managers and their incentives to develop effective strategies.

This chapter reviews the theoretical and empirical literature related to business strategy and corporate governance – two central issues when considering the competitiveness of firms. In this study we draw on March's exploitation–exploration framework to distinguish different approaches to formulating business strategy. It expands the framework using insights from institutional theory, the resource-based view and transaction cost economics. The discussion of corporate governance focuses on ownership structure and the boards. In addition, our study will extend the application of agency theory to corporate governance in an appeal to institutional theory. This discussion leads to our conceptual model. It provides a basis for identifying the need for further research and deducing detailed research questions to be addressed.

Key words: institution, business strategy, corporate governance, firm performance.

Introduction

As outlined in the introduction, there is a dual set of objectives in this study: the main empirical objective of this study is to analyse the business strategies and corporate governance adopted in Chinese firms, including those that are State-owned, collectively-owned and privately-owned. The theoretical objective is to integrate the perspectives of institutional theory, the resource-based view and transaction cost economics to explain exploitation and exploration strategy, and to explore the application of agency theory to discussing corporate governance in the Chinese context. Here we evaluate the relevant theoretical and empirical literature, of which there is a considerable body. It proposes an integrative framework that characterizes determinants for exploitation and exploration of strategic choices (institutional factors, resource and transaction cost) and corporate governance (ownership structure, board of directors and supervisory board) and the interrelationship implications within the Chinese social context.

This chapter is organized around six sections that reflect the empirical and theoretical objectives of this study. The next section critically reviews institutional factors related to business strategies and corporate governance. The third section discusses current theoretical knowledge on business strategies. Next, the theory related to corporate governance is analysed. The fifth section is an identification of the void in the literature. The final section concludes the chapter.

Institutional theory (IT)

Firms are not autonomous. They are embedded in a specific social and institutional context or institutional framework that both constrains and enables strategy and corporate governance (Morgan, 2004; Williamson, 1994). When formulating and implementing its business strategies, a firm needs to consider its external environment, including the broader institutional influences (Oliver, 1997). Institutional theory also is a lens to compare the corporate governance models (Nicita and Pagano, 2003). This is particularly important in transition economies, where firms are more susceptible to institutional influences and changes (Child, 1994; Henisz and Delios, 2002; Peng, 2003; Scott, 2002). Baron (1997) observes that since the business environment is composed of market and non-market components, any approach to strategy formulation must integrate both market and non-market considerations. The researchers who have investigated the strategy–structure–performance paradigm in the Chinese context generally conclude that performance is enhanced by the alignment of a firm's strategy and its environment (White and Xie, 2006; Tan, 2002; Tan and Tan, 2003, 2004). Understanding the institutional dimension of a firm enables us to better understand decisions made by firms about the use of existing resources, the acquisition or development of new resources, and where resources are developed internally, whether acquired on the market or from a membership of a network of aligned firms. Many authors observe that institutional factors modify the use of agency-theory hypotheses in transition economies (Bebchuk and Roe, 2004; Chen, 2005; Clarke, 2003). Institutional theory not only helps explain the diversity of ownership and that the board characteristics will persist due to the institutional differences such as culture, ideology, politics and legal origins, but also shows that institutional reform could provide incentives for the change of corporate governance.

Different institutional environments give rise to different business strategies and corporate governance practice. Institutional theory facilitates our understanding of the underlying determinant of the long-run performance (North, 1990). The framework of institutional theory is primarily concerned with an organization's relationship with the institutional environment (Martinez and Dacin, 1999). Institutions are the 'rules of the game' in a society, including formal rules (political rules, economic rules, and contracts) and informal rules (codes of conduct, norms of behaviour and conventions), which are embedded in culture and ideology (North, 1990). A social network based on interpersonal relationships is a particular type of informal institution that has long been important in China (Boisot and Child, 1996; Luo and Park, 2001; Peng, 2002, 2004). Informal institutions offer some constancy and predictability in the absence of well-developed formal institutions (Luo, 2003; Luo and Park, 2001; North, 1990). The formal and informal arrangements, after Scott (1995), possess regulative, normative, and cognitive pillars. The regulative pillar has to do with rule setting, monitoring and sanctioning activities. The normative pillar is a prescriptive, evaluative, and obligatory dimension. The cognitive pillar refers to the rules constituting the nature of reality and the frames through which meaning is conveyed (Scott, 1995). The formal institution is the regulative pillar while the informal institution combines the cognitive and normative pillars. As such, the institutional impact on a firm can be viewed from the formal and informal perspectives.

Firms as organizations are a response to the institutional contexts in a given economy (North, 1990; Oliver, 1991). The strategic choices of firms are selected within and constrained by the institutional frameworks (Peng and Heath, 1996). Different institutions can explain why the strategies of firms are not the same in different countries, regions or even at different periods in the same area (North, 1990). Differences in the country-level institutional environment may also explain the differences in corporate governance practices (Aguilera and Jackson, 2003; Gospel and Pendleton, 2003; Wittington and Mayer, 2000). When the institutional context changes we expect the strategies and corporate governance practices of the firm to adjust accordingly.

The economic reform process in transition economies has brought about huge institutional changes. The shift away from a command economy to a market-oriented economy implies that the institutions of the central planning regime have been replaced by more market-based institutions that facilitate economic exchange (Groves et al., 1994). While previous formal constraints of central planning have been weakened, the new formal institutions necessary for a market-based economy have yet to appear or are embryonic in form (Peng and Heath, 1996). Therefore, informal institutions such as networks that have been established during the past period of the planned economy continue to play a role regulating economic exchanges in these countries. Peng (2002) argues that there are three main informal institutions: interpersonal relations for managerial networking; external connections linking executives, key stakeholders, especially government officials; and the reputation of conglomerates. Informal institutional arrangements may remain operative despite official reforms because they are culturally based (North, 1990). This explains why some of the legal texts are similar between advanced countries and developing countries, but practices are so different (Gordon and Roe, 2004).

Institutional theory provides a framework enabling us to examine how institutional change facilitates and constrains a firm's strategic initiative and modes of corporate governance, which are the foci of the study. China is no exception to the observation that a firm's attributes alone cannot account entirely for variations in performance (Keister, 2000). With the economic reform in China, State-ownership and control have been gradually relaxed and public policies are more market-based (Chai, 1997; Chow, 2002; Garnaut et al., 2005). However, the growth of firms is still limited by formal institutions and cultural restraints (Child and Lu, 1996; Luo and Park, 2001; Peng, 2002; Peng and Heath, 1996). Therefore, this study considers the impact of the environment on business strategies and corporate governance from the perspectives of both formal institutions and informal institutions. Formal institutions mainly focus on policy, various legal texts and economic contracts. Informal institutions will emphasize socially sanctioned norms of practices such as networks, which are embedded in culture and ideology (Scott, 1995).

Business strategy (BS)

According to Baron (1995), a firm's business strategy comprises both market and non-market strategies. A market strategy is 'a concerted pattern of actions taken in the market environment to create value by improving economic performance', whereas a non-market strategy involves such actions as lobbying for policy changes favourable to the firm. The market orientation in business has been the main focus of the strategic management literature. The formation of a firm's strategies is dependent on the environment in which the firm operates (Gatignon and Xuereb, 1997; Kohli and Jaworski, 1990; Miller, 1988). The matching of strategy and environment can obtain better performance – a poor match can hurt performance (Grant, 1999; Miller, 1988). According to this perspective, strategy drives performance. A firm's business strategy must evolve dynamically to meet environment contingencies, in particular through reconfiguring its set of competences to sustain competitive advantage in a rapidly changing business environment (Eisenhardt and Martin, 2000; Karim and Mitchell, 2000; Teece et al., 1997).

Several typologies of strategic orientation have been set forth in the strategic management literature. Based on their field studies conducted in four industries (textbook publishing, electronics, food processing and health care), Miles and Snow (1978) proposed a strategic typology classifying business units into four distinct groups: prospectors, analysers, defenders and reactors. In 1980, Porter claimed that a company could follow only three generic strategies: a cost leadership strategy, a differentiation strategy and a focus strategy (Porter, 1980). March (1991) distinguishes two types of strategic orientation: an exploitation strategy and an exploration strategy. Generally the generic strategy typologies developed by strategy researchers have broad similarities (Campbell-Hunt, 2000). All include cost and efficiency as a key strategy (Thornhill and White, 2007). However, the original Miles and Snow research was limited in the number of industries and the range of capabilities studied (Desarbo et al., 2005). They did not attempt to prove the validity of their typology across other industry types. Porter's (1980) scheme is criticized as it 'is described in relatively general terms, and seems to be limited explaining the competitive market behavior of larger firms' (Smith et al., 1989: 63). For the purpose of this study, we apply the March's exploitation–exploration strategy framework to investigate the different strategic orientations that firms may adopt. The concepts of exploitation and exploration learning (March, 1991) underpin organizational adaptation research (Gupta et al., 2006).

Firms that wish to reconfigure their competence base can deploy an exploitation or exploration approach, or a combination of both. Exploitation involves 'co-ordinating, preserving and supervising' the use of existing resources and thus leads to the improvement of current (short-term) performance, whereas exploration is about identifying opportunities and taking the firm in new directions with new capabilities and thus leads to the improvement of future (long-term) performance (March, 1991). The challenge for managers is to decide which approach to take to realize beneficial strategic change in response to change in the business environment. Integration of the exploitation–exploration framework with the institution theory enables us to examine the dynamic strategic fit of the firms in a transition environment.

Since concepts of exploitation and exploration are analogous to those of asset deepening and asset extension in the literature of strategic management, the following sections discuss the theories related to these concepts (Karim and Mitchell, 2000). In addition, exploration strategies should take into account transaction costs on the demand side in the context of the choices between 'make' and 'buy' (Foss and Foss, 2006). These costs are associated with search, screening, processing and contracting (Antonelli, 2004). This study has drawn on two perspectives in order to explain exploitation and exploration strategies: resource-based view and transaction cost economics. A firm's strategy selection is based on the careful evaluation of its resource portfolios (Barney, 1991). There are certain 'blind spots' in this thinking. It has not adequately addressed how firms acquire or create new resources. Starting with whatever initial complement of resources and capabilities it had at its founding, a firm subsequently has to acquire the additional resources and capabilities to compete and survive. This can be done through a combination of make, buy and ally, in which case it trades off various forms of transactions costs associated with buying in what it needs on the market, making it internally or acquiring its needs using some form of intermediate contract such as a joint venture or alliance (Williamson, 1975).

Resource-based view (RBV)

The discussion of business strategy in this study begins with a theoretical review of the resource-based view (RBV) of the firm. Strategy is the fit between the firm's external situation and its internal resources and capabilities (Grant, 1991). Significant changes in the business environment will require organizational changes – different resources and capabilities – in order to realign the firm with its environment. A RBV perspective focuses inwardly on the firm's resources and capabilities to enhance its competitive advantage (Barney, 1991; Penrose, 1959; Peteraf, 1993). The RBV approach helps us understand how firms achieve and sustain competitive advantage through resource building as well as leveraging the existing resources.

A firm is a bundle of resources and routines that influence growth (Barney, 1991). From the resource-based perspective, a firm's competitive advantage comes from its superior resources. A firm is therefore advised to choose its strategy based on its resources (Barney, 1991; Penrose, 1959; Peteraf, 1993). According to the RBV, only the 'valuable, rare, imperfectly imitable and non-substitutable' resources are sources of the sustainable competitive advantage (Barney, 1991). The underlying assumption is that unique sets of resources and associated capabilities protect the firm from imitation by competitors and provide the basis for an accumulation of superior profits through differentiation of its product and services (Porter, 1986). Unique resources and capabilities ensure profits, expand business opportunities and improve the performance of the firm. Resources can be classified into physical resources such as plant and machinery, human resources such as managerial and technical staff, and organizational resources such as information and experience.

Possessing unique resources is not sufficient for competitive advantage. A firm needs to have the capability to deploy those resources. According to Grant (1991), a capability is a set of skills vested in people or the organization that enables the firm to deploy the asset to achieve the competitive advantage. 'While resources are the source of a firm's capabilities, capabilities are the main source of its competitive advantage'

(Grant, 1991: 119). The link between a firm's resources and capabilities and the competitive advantage of a firm remains a central concern of the scholars in the area of strategic management (Barney, 1991, 2001; Grant, 1991; Leask, 2004; Peteraf, 1993; Porter, 1986; Wernerfelt, 1984). Competitive advantage is at the heart of the strategy (Porter, 1985), based on the resources and capabilities available to the firm. Resource-side issues have to be dealt with by the practising strategists (Priem and Butler, 2001b). A firm is said to have a sustained competitive advantage when it outperforms its competitors by deploying firm resources more cost efficiently or more distinctively (or both) than its competitors (Barney, 1991; Porter, 1985).

Differences in the underlying resources of the firms enable them to engage in some competitive actions and not others. If a firm owns the resources that are 'strongly shielded' from competitive pressures, its strategies will mainly focus on how to maintain, strengthen and augment the existing competitive resources (Williams, 1992). These kinds of resources include higher order, complex and intangible resources such as knowledge, patents and skills and competencies. If the resources of a firm are easily transferred or replicated, such as cheap labour or imitable technology, there are several avenues for it along which to compete. It may pursue a strategy that is proactive, risk taking and innovative instead of a cost leadership or differentiation strategy (Miller, 1983). Alternatively, the firm adopts a strategy of short-term harvesting or it invests in acquiring new sources for competitive advantage (Grant, 1991). In order to upgrade a firm's competitive advantage and broaden the firm's strategic opportunity set, the resource gap in the firm must be filled (Grant, 1991).

Although some resources and capabilities are standard across all economies, others are especially prominent in emerging economies (Hoskisson et al., 2000). This is because the value of particular resources depends on the specific market context in which they are applied (Barney, 2001). Priem and Butler (2001a) note that greater efforts should be made to establish appropriate contexts for the RBV. In emerging economies, resources that are valuable in a market context such as managerial expertise and financial resources are likely to be scarce because they had not been fostered under the previous State-organized economic system. In addition, a good relationship with the government is considered to be a crucial resource in these countries (Hoskisson et al., 2000; Peng, 2005).

Moreover, according to RBV, the strategic decisions of firms will be shaped by the cognitive capabilities of their managers (Wright et al., 2005). Strategic restructuring may be impeded not only by constraints related to the lack of organizational resources, but also by a lack of managerial ability to undertake change (Mahoney, 1995). Strategic flexibility is the joint outcome of a firm's resources and its ability to coordinate their uses. Managers' flexibility in reconfiguring, developing and using resources are the most critical tasks in distinguishing successful from unsuccessful firms in emerging economies (Uhlenbruck et al., 2003).

RBV thus enables us to identify two dimensions of business strategy useful for this study. One has to do with identifying the existing resources and capabilities of the firm, while the other deals with identifying the resources needed for the growth of the firm. That some firms are able to survive and compete despite fierce competition indicates that they possess certain advantageous resources and capabilities (Bruton et al., 2000). Identifying these resources and capabilities helps the managers of the firm to maintain their current competitive advantage. As a competitive market develops, the development of new resources becomes more important (Hoskisson et al., 2000). Developing new resources helps managers improve the competitive advantage of their firm and expand business opportunities.

However, RBV is criticized for being overly narrow at its analytical core, because it argues that all performance differences are explicable in terms of the differential efficiency of the resources underlying strategies (Foss, 2002). Often the firm achieves a sustainable competitive advantage because it reduces opportunistic behaviour and allows for firm-specific investment (Mahoney, 2001). In addition, RBV is criticized for failing to answer questions about how the resources can be obtained (Priem and Butler, 2001a). Firms must create new resources and capabilities if they want to survive in the long run. New resources and capabilities can be acquired through development by investing internally, development through external governance such as purchase from the market, or development through merger and acquisition (M&A) and development through a networking structure (Penrose, 1959; Williamson, 1985; Luo, 2003). The issue of strategic choices on how to seek previously unavailable resources is especially pertinent to transition economies.

At this point, it is possible to take a broader view to address the key issues of strategy, while still keeping the efficiency perspective characteristics of the RBV (Foss and Mahnke, 2002). Transaction cost notions are useful for developing such a broader view.

Transaction cost economics (TCE)

Significant changes in the environment will likely require different or new resources and capabilities in order to realign the organization with its environment (White and Xie, 2006). There exists an extensive literature to explain business strategies from both the RBV and TCE approaches (Foss, 2002). TCE helps to explain efficiency-based sources of competitive advantage such as superior organizational arrangements. Since transaction costs in transition countries are much higher than those in advanced economies (Meyer, 2001), the consideration of strategic choices from the perspective of TCE as well as RBV will be more complete.

Transaction costs are present in each organizational governance form (Coase, 1937), no matter whether the transaction is conducted on the market, within the firm or via an intermediate form such as an alliance or partnership (Williamson, 1975, 1985). Williamson (1975, 1985) observes that transaction costs are the outcome of bounded rationality (cognitive and language limits on individuals' abilities to process and act on information), opportunism (self-interest seeking with guile) and asset specificity (the dedicated assets with respect to use or users, which implies sunk costs).

The cost of using the market to acquire resources and other inputs is the cost of using the price mechanism (Coase, 1937). In transition economies, transaction costs are high because the price system often does not accurately provide signals for efficient resource allocation (Hoskisson et al., 2000). The cost of transacting on the market generates a preference for hierarchical governance – the internalization of transactions within the firm. Problems associated with bounded rationality, opportunism and asset specificity are reduced through organizing the activity in the firm (Williamson, 1985).

Firms seek internalization of transactions such as mergers and acquisitions (M&A). Mergers and acquisitions aim to avoid the problem of asset specificity in maintaining a transactional relationship with another firm. In transition economies, however, the lack of strategic factor markets such as financial markets, unclear property rights, and an inadequate legal framework to safeguard the market actively raise the transaction costs of market-based M&As (Peng, 2002; Peng and Heath, 1996). Some M&As in emerging economies are ordered by the government, which wishes the well-performing businesses to acquire poorly performing businesses (Li and Wong, 2003). Peng and Heath (1996) suggest that hybrid structures such as managerial networking are more efficient solutions in transition countries. The parties of the network can secure credible commitment for sharing the resources instead of searching for them in the market (Boisot and Child, 1996; Luo, 2003; Peng and Luo, 2000). This networking is used to reduce transaction costs by establishing credible commitment in an institutional environment characterized by weak property rights and irregular monitoring and enforcement of formal rules by the central State (Nee, 1996; Peng and Heath, 1996; Peng and Luo, 2000). The success of this network is a product of both partners achieving their goals (Ireland et al., 2002). However, in an uncertain environment, opportunistic behaviour, divergent economic interests and learning barriers including cultural barriers and absorptive capacity of the acquired resources through a network are the costs inherent in the network (Peng, 2002).

Resource allocation is dependent on the means for the economization of the transaction costs (Williamson, 1994). When the cost of using the market is greater than the cost of organizing within the firm, the allocation of resources and the transaction will be implemented in the firm or via the adoption of intermediate forms. Similarly, when the costs (coordination costs) of organizing within the firm rise, a firm might revert to the market (outsource, contract or direct purchase). The make-or-buy decisions are an outcome of managers' simultaneous consideration of elements of the firm's external and internal environment (White, 2000). It is a manager's responsibility to form an overall assessment of the situation on the basis of a costs and benefits evaluation of the strategies (Lockett and Thompson, 2001).

A classical analysis of transaction cost is based on the situation in the United States (Boisot and Child, 1988). There is little discussion of firms' strategies from the TCE perspective in transition countries (Hoskisson et al., 2000). Extending the existing framework and applying these concepts to an analysis of transition countries is essential (Boisot and Child, 1988). This study will contribute by discussing the link between business strategies and transaction costs within the context of transition economies. Because transition countries have higher transaction costs owing to the institutional deficiencies (North, 1990), adopting a strategy that decreases transaction costs is important for managers. This suggests that comparative research of the different modes of governance needs to be carried out in the study. Based on the comparison, an appropriate strategy for new resource exploration will be suggested.

Exploitation and exploration strategies

All firms face the same organizational challenges: starting with whatever initial complement of resources and capabilities they had at their founding, they have to acquire the additional resources and capabilities that enabled them to compete and survive, through a combination of make, buy and ally strategies. Two distinct ways of developing the resource and capability base of the firm are exploitation and exploration strategies (March, 1991). Surviving in a transition economy often requires that firms pursue a dual strategy that attempts to balance the need to leverage their current competencies and resources, which is a strategy of exploitation, while preparing to acquire resources anticipated for the future through a strategy of exploration and experimentation (Wiseman et al., 2006). Both activities consume scarce resources, which require organizations to set explicit decision-making policies for allocating the resources available for resource renewal (March, 1991).

Exploitation is a strategic renewal process aimed at leveraging existing firm-specific assets by improving them or by improving their use. It includes matters such as refinement, choice, selection, efficiency, implementation, focused attention, and developing reliability in experience, and it is related to the use of existing resources and capabilities (Crossan et al., 1999; Hitt et al., 2002; Holmqvist, 2004; Levinthal and March, 1993; March, 1991). Exploration, by way of contrast, is a strategic renewal process that seeks to acquire new firm-specific assets. Exploration strategies are those that explore new domains, with the intention of acquiring new resources and capabilities (but for which the probability distribution is not known) (March, 1991). Exploration represents a combination of search, innovation, variation, experimentation, trial and free discovery, and is concerned with variety in experience (Holmqvist, 2004; Levinthal and March, 1993; March, 1991). Based on these two definitions, exploitation is a requirement for implementing an advantage-seeking growth strategy, and exploration is needed for succeeding in opportunity-seeking growth (Caldart and Ricart, 2007). In addition, this study also refers to the definition in Meyer's (2007: 1500) study of exploitation vs exploration in transition economies: 'Exploitation learning refers to the pursuit and acquisition of knowledge, which is new for the companies in a transition economy, but already in existence in the West. Exploration learning is the creation of new knowledge to develop strategic flexibility, leading to sustainable competitive advantage.'

Two competing views of research have been proposed concerning the choices between exploitation and exploration: punctuated equilibrium versus ambidexterity (Gupta et al., 2006; Kyriakopoulos and Moorman, 2004). The first is based on a belief that exploitation and exploration are separate, mutually independent activities, and exploitation precludes exploration (Weick and Westley, 1996). Those who hold this view explain it in three aspects. To begin with, as both activities compete for scarce resources, exploitation strategies may limit the amount of firm exploration and vice versa (March, 1991). Second, exploitation and exploration strategies are associated with opposite organizational structures and cultures. Firms that pursue both strategies are viewed as lacking focus and internal fit (Miller and Friesen, 1986). Third, firms should utilize one of these strategy approaches to optimize fit with the external environment including infrastructure, economy and market conditions (Galbraith, 1973; Lawrence and Lorsch, 1967). For example, in a highly turbulent market, increasing investments in the exploitation of existing processes is proportionally more advantageous than in stable environments. This is because turbulent environments require continuous adaptation (Luo, 2000; Masini et al., 2004; Zollo and Winter, 2001). On the other hand, some researchers argue that exploitation may be appropriate when the environment is stable and a firm's initial set of resources and capabilities is adapted to that environment (Isobe et al., 2004). When the environment undergoes major changes such as in China, the firm must be able to continually explore new resources, capabilities and structures that better match the new environment (Teece et al., 1997; White and Xie, 2006).

Proponents of ambidexterity view exploitation and exploration as complementary, despite the acknowledged tension between the two practices. Although exploration primarily involves the acquisition of new knowledge from external sources, it may involve the novel combination of existing technologies and know-how. According to Levinthal and March (1993: 105), firms must engage in both strategies: 'An organization that engages exclusively in exploitation will ordinarily suffer from obsolescence. The basic problem confronting an organization is to engage in sufficient exploitation to ensure its current viability and, at the same time, to devote enough energy to exploration to ensure its future viability.' A balanced focus would appear more suitable for businesses operating in complex environments such as a transition economy.

Corporate governance (CG)

Corporate governance provides an organizational context within which managers adjust their product market strategies to compete (Bower, 1970). Corporate governance is an integrated set of internal and external controls that harmonize manager–shareholder conflicts of interest resulting from the separation of ownership and control (Williamson, 1984).

There are several models describing the nature of corporate governance that have gained predominance globally. The most popular classification is the Anglo-Saxon model versus the Rhineland (or german–Japan) model. The Anglo-Saxon model applies to the UK and the US, and the

Rhineland model applies to Germany and some continental European countries (Becht and Roël, 1999; Fukao, 1995; La Porta et al., 1998; Shleifer and Vishny, 1997). Corporate governance in Japan is seen as an asian variant of the Rhineland model. The Anglo-Saxon model of governance is characterized by widely held equity and a focus on financial objectives. The driving force of this model is the external market system. The capital market is efficient. Outside directors represent about 75 per cent of the board in large publicly held companies in the US and 50 per cent in the UK (Hovey, 2005). The primary objective of this governance system is to represent the shareholders and to maximize shareholder wealth (Monks, 2001; Salmon, 2000). The Anglo-Saxon model relies on the consistent functioning of unitary boards and there are no supervisory boards (Monks, 2001). The power of employees is brought to a minimum, institutional portfolio investors are powerful, capital markets are strong and takeover activities are common (Prigge, 1998).

The Rhineland model is characterized by a significant holding by a parent company, and outside shareholders represent a smaller portion of the equity (Cheung and Chan, 2004). A successful example of Rhineland system is used in Germany. In the Rhineland model, individual companies within a particular company group can be viewed as an 'internal market', both in terms of financial and other resources such as labour and intellectual property (Cheung and Chan, 2004). This is why the Rhineland model was called an insider, networked, bank-based or closed system; in contrast to the open, market-based Anglo-Saxon model (Prigge, 1998). The Rhineland model is also characterized by a two-tier board structure. German companies, for instance, comprise a two-tier board system in which responsibilities are split between the executive board and a supervisory board. Many have employee representatives on supervisory boards (e.g. in Scandinavia), and in many countries banks are highly involved in control of the corporate sector (e.g. Germany and Italy) (Wymeersch, 1998).

Starting from the seminal work of Berle and Means (1932), analysis of corporate governance deals with different mechanisms such as the rights and equitable treatment of shareholders; the role of stakeholders in corporate governance; disclosure and transparency; and the responsibilities of the board (OECD, 2004). According to Turnbull (1997), the primary methods of analysing corporate governance are the principal–agent perspective, the stakeholder perspective and the political perspective. The principal–agent perspective recognizes that the central problem in corporate governance is to construct rules and incentives to effectively align the behaviour of managers (agents) with the desires of principals (owners) (Hawley and Williams, 1996: 21). The stakeholder perspective views the firm as comprising a system of stakeholders operating within the larger system of the host society that provides the necessary legal and market infrastructure for the firm's activities. The political model recognizes that the allocation of corporate power, privileges and profits between owners, managers and other stakeholders is determined by how governments favour their various constituencies (Turnbull, 1997). Analysis of corporate governance should focus on different perspectives in different countries because of their distinct circumstances at the time or historical accidents (Bebchuk and Roe, 2004).

The major issue for China's corporate governance is the high agency cost associated with severe conflict between principals and agents, which is brought about by the special ownership structure of China's listed firms (Tam, 1999; Tenev et al., 2002). For the purpose of this book, the principal–agent perspective is adopted. This study focuses on ownership structure and the boards because they are two crucial internal governance factors in resolving the conflicts between owners, in particular minority shareholders and managers in order to reduce agency costs (Denis and Kruse, 2000; Liu, 2006). Ownership structure is crucial to align the interests between shareholders and managers, while the board can exert influence on the behaviour of managers to ensure that the company is run in their interests (Barnhart et al., 1994). These two issues are discussed below.

Agency theory

Agency theory has become the dominant theoretical framework in English-language corporate governance studies (Shleifer and Vishny, 1997). Berle and Means (1932) argue that the agency problem stems from the separation of ownership and control in modern corporations, which gives rise to information asymmetry between managers and the shareholders. The managers (agents) possess more expertise than the shareholders (principals), thus giving them more latitude for self-interested behaviour (Shleifer and Vishny, 1997). The goals the agents pursue are therefore sometimes not aligned with those of the principals. The costs resulting from managers misusing their position, as well as the costs of monitoring and disciplining them to try to prevent abuse, have been called 'agency costs' (Blair, 1995). As a consequence, one of the important mechanisms of corporate governance is control of the agency problem in order to increase productivity and managerial efficiency (Fama and Jensen, 1983).

Ownership structure and the boards are two crucial internal governance factors in resolving conflicts between owners and managers (Denis and Kruse, 2000; Liu, 2006). Ownership structure is instrumental in aligning the interests between shareholders and managers, while the board can exert influence on the behaviour of managers to ensure that the company is run in their interests. These two issues are discussed below.

Ownership structure

Jensen and Meckling (1976) delineate a difference between 'capital structure' and 'ownership structure'. The former usually refers to the relative quantities of bonds, equity, warrants and trade credit, which represent the liabilities of a firm in a market economy. The latter refers to the relative amounts of ownership claims held by insiders (management) and outsiders (investors with no direct role in the management). This study uses the latter definition.

Ownership structure involves many dimensions, among which the most important are the allocations of residual control rights and rights to residual benefits. An ownership structure that is consistent with the objective of firm-value maximization may require that the residual claimants, who contract for the residual benefits, bear the residual risks, the 'risk of the difference between stochastic inflows of resources and promised payments to agents' (Fama and Jensen, 1983: 302). Ownership structure needs to be organized so as to provide an incentive mechanism that aligns the interests of decision makers (managers, contractual partners) and those who own assets, and motivate the agents to take appropriate actions, thus ensuring the efficient allocation of resources (Zhou, 2001).

Franks and Mayer (1992) distinguish between two broad categories of corporate ownership structure across countries. In the first category are the countries of continental Europe and Japan, in which the ownership of individual firms is often concentrated within a small number of other directly related firms, banks and families. In the second category, which includes Britain and the United States, ownership is dispersed among a large number of unrelated individual and institutional investors, and cross-shareholdings are rare. Owner categories such as the State, companies, families, institutions and banks are found to differ substantially in terms of the largest owner's share. The State, companies and families have a preference for control and majority ownership (Thomsen and Pedersen, 2000).

There has been a vigorous debate on the importance of the distribution of share ownership for firms, especially that of State ownership. Research on this issue has yielded conflicting results. One key debate is whether State ownership is generally associated with inferior performance (e.g. Hart et al., 1997; Oswald and Jahera, 1991; Pfeffer and Salanick, 1978; Shepherd, 1988; Shleifer and Vishny, 1997). There are two main reasons for believing that State ownership impairs performance. The first one has to do with the non-profit-maximizing behaviour of State ownership. Governments are more interested in realizing low output prices, employment or external effects than in profitability and overcoming the market failures associated with private firms (Hart et al., 1997; Shepherd, 1988; Thomsen and Pedersen, 2000). The second reason is that State ownership is considered to possess significant agency costs. The de facto absence of owners in firms increases managerial discretion in a potentially adverse way, because the monitoring of managers is much more difficult in SOEs (Agrawal and Knoeber, 1996; Jensen, 1986; Jensen and Meckling, 1976). Thus SOEs are incompatible with the managerial structure of modern firms in terms of ownership efficiency.

In contrast, privately-owned firms are considered to be more successful than State-owned firms in addressing problems of corporate governance (Hu et al., 2004). Kornai (1990) develops his critique of the State socialist economy largely from the vantage point of property rights as an incentive mechanism. The State ownership of firms generated soft budget constraints, which are said to exist whenever a loss-making company continues to receive financing (Kornai, 1979, 1980). A soft budget constraint alters the incentive environment in which managers operate: they are free to pursue goals like prestige and promotion without worrying about profitability. The soft budget constraints provide distorted incentives for State firms, inducing inefficient allocation of resources and economic behaviour. Privately owned firms, by contrast, face hard budget constraints, which subject firms to the discipline of the marketplace and limit the politicization of the economy and the investment decisions (Kornai, 1979, 1980). In this light, then, private property rights can more effectively align the incentives provided to the managers and the expected gains for the owners. Moreover, Kornai (1990) sees private ownership as an effective means of resisting the intervention of the political authorities in transition economies. He states plainly: 'It is futile to expect that the State unit will behave as if it were privately owned and will spontaneously act as if it were a market-oriented agent' (Kornai, 1990: 58). This perspective has been widely influential in views expressed about the transformation of State socialist economies.

On the other hand, however, there is considerable evidence that the distribution of ownership does not necessarily have an influence on performance (Linz, 1997; McDonald, 1993; Whitley and Czaban, 1998). Most of the studies providing evidence for the relationship between ownership and performance rely on the assumption of exogeneity, which means that ownership is external or outside the nature of the enterprise (Goergen, 1998). Yet, the relationship between ownership structure and firm performance is insignificant when controlling for endogeneity of ownership structure (Demsetz and Lehn, 1985; Demsetz and Villalonga, 2001). The endogeneity problem arises when ownership is chosen as a function of performance or as a function of unobserved variables that also affect performance. In other words, ownership structure has been justified in terms of a series of factors within the firm itself, inherent to the area of industry or sector in which it operates (Leech and Leahy, 1991). Research on ownership endogeneity concludes that ownership is not due to value maximizing behaviour, but rather determined by the circumstances or factors of the firm such as its contracting environment, size, the inherent riskiness of the assets, or its performance (Demsetz and Lehn, 1985). It is unreasonable to suppose that ownership structure has an impact on profit maximization (Demsetz, 1983).

In addition, agency problems arise in privately owned firms as well as State-owned ones (Vickers and Yarrow, 1991). The privately owned firms are not necessarily better governed (Chang and Singh, 1997). Neither has a perfect disciplinary mechanism. Privately owned firms therefore may perform worse than SOEs if their corporate governance is badly flawed. Boardman and Vining (1989) further point out that the weakness of the agency theory above is that it is based on North American firms. There is no inherent reason that public enterprise led by government officials cannot achieve the high levels of economic performance of a private enterprise economy (Walder, 1995b). Using data based on firms in Europe, Japan and Canada, SOEs are not by definition inefficient (Boardman and Vining, 1989; Walder, 1995a, 1995b).

The advantage and disadvantage of any form of ownership is a pragmatic issue. Different forms of ownership are appropriate to countries at different development levels and reflect markets of varying degrees of maturity (Jiang, 2001c). Moreover, ownership structures are not the same in different industries. The characteristics of an industry, including the intensity of competition and maturity, directly affect the corporate ownership structures found among firms. The frequency of government and company ownership accordingly vary widely around the world (Thomsen and Pedersen, 2000).

The characteristics of the board

Another factor that resolves principal–agent conflict is the organizational form of the board (Barnhart et al., 1994). Agency theory examines board effectiveness based on the assumption of 'goal conflict' between principal and agent (Eisenhardt, 1989). Agency theorists view the board as critical in situations where agency costs between shareholders and managers are severe (Barnhart et al., 1994; Williamson, 1984).

The board has the legal authority to ratify and monitor managerial initiation, and evaluate and reward, or penalize, the performance of top managers. By performing its control role, a board attempts to align the interests of senior executives and those of shareholders to minimize agency costs. The board's choice of managerial controls can influence the decisions of top managers and, ultimately, the direction of corporate strategy (Beekun et al., 1998). Changes in board characteristics may change the relationship between top managers and shareholders (Baysinger and Hoskisson, 1990). Since the board's characteristics, such as its composition, are related to corporate strategies, secular changes in board composition, through changes in the emphasis of control, may have important strategic implications for the corporation (Baysinger and Hoskisson, 1990).

Pistor (1999) defines two board structures: one is the unitary board, which means that corporations are run under the supervision of a single board; the other is the two-tier board found in some European countries, in which the board is divided into an executive board and a supervisory board. Different models of board structure have been formed in different economic and social systems. Unitary boards can be found in countries such as the US and the UK, while in countries such as Germany, Holland and Austria the two-tier board system has been adopted. The two-tier board is also the type found in China (Tam, 1990; Carrasco, 2005).

In a unitary board, the board of directors must serve the functions of allowing managers the flexibility they need to run the business while setting policy and appointing the principal senior executives, selecting the officers who in reality manage the business, and monitoring the managers to limit self-dealing and poor management (Johnson et al., 1996). In the paradigm countries such as the US and the UK, the operation of the external market system is efficient, thus the monitoring of the BoD is external, through the operations of the market for corporate control (Shleifer and Vishny, 1997; Roe, 1997). The market system, combined with an active board, is demonstrated to be a relatively effective instrument for reducing agency costs (Hovey, 2005).

The two-tier board system is a mechanism in which responsibilities are split between the executive board and the supervisory board. The executive board is in charge of the daily management of the company's business, while the supervisory board advises and monitors the performance of the executive board (Charkham, 1995; Charles, 1994; Dahya et al., 2003; Tam, 1995). The aim of the two-tier board system in Germany is the promotion of trust, cooperation and harmony. As a kind of continuous representative of the shareholders between their meetings, the supervisory board is supposed to be the guardian of their interests (Dahya et al., 2003). In some European countries such as Germany, the supervisory board is mainly composed of capital representatives who act on behalf of the shareholders and creditors, and labour delegates who look after the interests of the employees (Tenev et al., 2002).

There has been debate over the pros and cons of the unitary board vis-à-vis the two-tier board structure. The combination of supervisory and management function in the unitary board system suggests the advantage of no real conflict between the shareholders and the board (Schneider-Lenne, 1992). The unitary board may result in a closer relationship and better information flow between the supervisory body and managerial bodies (Denis and McConnell, 2003; Roe, 1997). However, unitary control occurs when a single board governs a firm, which brings about four inherent problems: a) the corruption of power, b) lack of independent feedback, c) loss of information and bias and d) information overload (Turnbull, 1994, 2000). Thus the system lacks the level of checks and balances that the two-tier board system has (Charkham, 1995; Franks, 2000; Franks and Mayer, 2001). Such issues can be resolved by separating the supervision and management function as characterized by the two-tier system, in which a clearer and formal separation between the supervisory body and those being 'supervised' is made (Charles, 1994; Davidson, 1994). The proponents observe that this system provides a relatively strong sense of codetermination, which is a useful supplement to traditional contractual arrangements (Tenev et al., 2002). As such, the supervisory board will be responsible for supervising and ensuring that the executive board establishes the company's goals on behalf of the shareholders, and identifying the common interests of shareholders (Davidson, 1994). However, recent studies of firms using the two-tier board system find that this system could result in a contradiction of interests (Gorton and Schmid 2000; Tüngler, 2000). Critics point out that the two-tier board system empowers employees and that they use this power in ways that contradict the desires of shareholders, that is, they change the firm's objective function (Tenev et al., 2002;

Tüngler, 2000). Therefore, the efficiency of the two-tier board structure is undermined.

A company's ownership structure or board characteristics at any time may depend partly on the patterns it had earlier (Bebchuk and Roe, 2004). Put another way, its characteristics are path-dependent or institutionally constrained. Due to sunk adaptive costs, network externalities, complementarities, and the relative efficiency of alternative ownership structures, characteristics of the board depend partly on the structures with which the company and/or other companies in its environment started (Bebchuk and Roe, 2004). Models of corporate governance must go beyond the traditional agency cost theories to include institutional theory in transition economies (Nicita and Pagano, 2003). Transplanting some of the formal elements of corporate governance from other countries without regard for the institutional complements may lead to serious problems later (Gordon and Roe, 2004).

Based on the literature, the research model is illustrated in Figure 2.1. The model aims to explore the interrelationship between business strategies, corporate governance, and the competitive position of the firm. Business strategy is discussed from the perspective of the exploitation vs exploration strategy. It tackles the issue of the relationship between strategies and the competitive position. Institutions, resources and transaction costs are three factors involved in discussing business strategy.

Figure 2.1 Theoretical model

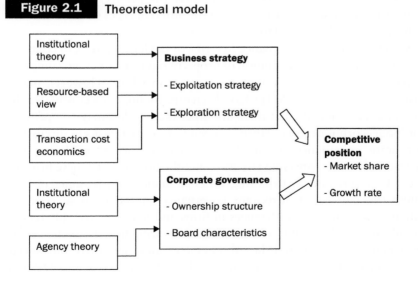

Business strategy is a necessary but not sufficient condition for a competitive position. A competitive position is also influenced by the incentives given to managers to adopt appropriate strategies (Filatotchev and Toms, 2003) and by the networks that tie together firms and managers (Boisot and Child, 1996; Peng, 2002; Peng and Heath, 1996; Luo and Park, 2001; Luo, 2003). Corporate governance tackles the influence on the competitive position of two factors: the ownership structure and the board characteristics. Moreover, the relationship between business strategies and corporate governance is also discussed in this study. The competitive position is measured by market share and the growth rate of the annual sales of the firms' main business.

Gaps in the literature

The preceding literature review focuses on two major substantive (theoretical and empirical) issues. There are a number of concerns in current research on transition economies in general, which this book would like to address.

First, with regard to business strategies, the literature is relatively sparse and lacks a unifying framework for Chinese firms (Tsui et al., 2004). Although some studies look at the business strategies from the RBV, the TCE and the IT respectively (Barney, 1991; Tan and Litschert, 1994; Tan, 1996; White, 2000), this study intends to move one step further by attempting to combine the three perspectives. The questions that naturally arise are whether or not the different types of Chinese firms behave differently or similarly in the market.

Second, corporate governance in terms of the legal system, in contract enforcement and in the definition of property rights in advanced countries is hardly a model of corporate governance in China. However, many firms in China including SOEs have been outstandingly successful (Singh, 2002), which suggests that there are other factors that may compensate for 'poor' corporate governance. The study faces two tasks in discussing corporate governance in the Chinese CE sector: a) analysing the characteristics of ownership structure and the boards of Chinese firms in the CE sector; and b) exploring different underlying parameters that affect the pattern of corporate governance in the Chinese context: differences in opinion, differences in firms and markets, and differences in culture, ideology and politics. This study will show how choices of ownership structure and board characteristics adapt to the social environment. It aims to examine whether SOEs, COEs and DPOEs differ dramatically from each other in terms of corporate governance structure. In other words, this study tries to address the following issues: do listed SOEs suffer more from corporate governance problems than non-SOEs? If they do, in what corporate governance mechanisms are the SOEs lagging behind non State-owned firms?

Third, the integration of business strategy and corporate governance is relatively scarce. This study integrates business strategies and corporate governance. Extant research has largely addressed the relationship between business strategies and a firm's financial performance, and the relationship between ownership type and a firm's financial performance, with little attention to the means of achieving a higher performance. However, 'financial performance' dimensions offer an incomplete or even misleading picture and are heavily constrained by pre-existing institutional arrangements.

Fourth, as regards methodology, most research results published by Chinese scholars are supported by systematic statistics (Jiang, 2001c). However, these research results are based on aggregate figures for various industries. Such figures help clarify the general transformation but limit the insights for individual firms. Moreover, these researches have neglected an important question: are there striking differences between industries in terms of the reform process and characteristics? If the differences between industrial sectors are large, 'then they are nothing but a group of vastly different average figures that can hardly represent the general feature of the reform in general' (Jiang, 2001c: 4). As a consequence, there is a need for further research using case studies that focus mainly on the same industry.

This analysis of a gap in the literature has identified the need for further research. This study aims to address these open issues: integrating IT, RBV and TCE to discuss exploitation and exploration strategies, analysing ownership structure and the boards of listed firms of a particular industrial sector in the Chinese context, and the integrating business strategy and corporate governance to examine the development model of firms in China.

Conclusion

In this chapter, the literature was reviewed critically for two distinctive research areas: business strategy and corporate governance. There is a considerable body of literature in relation to many aspects of business strategy and corporate governance. This chapter has brought together a review and analysis of the relevant literature for this study. The ultimate aim of this review must be to examine these areas in relative detail and apply this to an understanding of possible issues in business strategies and corporate governance, and policy advice for China.

The relevant business strategy literature was reviewed from the perspectives of exploitation vs exploration strategies integrating IT, RBV and TCE. The relevant corporate governance literature was reviewed with a particular focus on ownership structure and the board. Based on this review, the authors developed a research model as shown in Figure 2.1.

3

The economic background in China

China's progress during the economic reform era that began in 1978
has been one of the great economic success stories of the post-war era.
(OECD)

Abstract: The various institutions that surround organizations – the
organizational environment in which firms are embedded – shape
the actions of firms in a number of subtle but substantive ways.
Given the influence of the environmental context on firm behaviour,
any choice of strategy or corporate governance that a firm makes is
inevitably affected by the constraints of its institutional environment.
This is particularly important in transition economies, where firms
are more susceptible to strong State-related institutional influences.
This chapter shows how the environment of China's economy
evolves. In this section we discuss economic reform in China, the four
stages or periods of development in the Chinese consumer electronics
sector, the development of firms with different ownership models and
the shareholding system in China, and we describe the case study
firms we have focused on in this study. The analysis indicates the
complexity of the context in which Chinese firms operate and
provides for a broad understanding of the economic background,
evolution and current state of enterprise reform in China.

Key words: economic reform, firm development, ownership, China.

Introduction

This chapter presents a review of the overall economic setting in China
and the evolution of economic reforms in China. Chinese economic

reform over the past three decades has succeeded in creating more competitive markets for all types of firms than in the pre-reform period. The State-dominated mechanisms that determined Chinese growth in the past have lost much of their influence. An increase in the number of firms and competitive intensity has transformed the motivation of the firms from accommodating the dictates of economic planners to meeting the demands of the market. Throughout the reform era, the realization of Chinese economic potential has rested on continuing and strengthening domestic economic reforms (Naughton, 1995).

This chapter begins with an overview of economic reform in China. Next, the development of reform centred on the growth of the Chinese consumer electronics sector is examined, followed by an analysis of the development of Chinese firms with different ownership. The fifth part discusses reforms related to shareholding forms of corporate structures in China. Part six provides a background to the cases including profiles of the cases and their respective competitive positions. The chapter concludes that institutional arrangements under the present economic reforms have been generally conducive to sustained economic growth.

Economic reform in China

China has achieved notable economic growth since the late 1970s (see Figure 3.1). Overall economic growth has been outstanding, with real gross domestic production (GDP) growth averaging almost 10 per cent since the 1970s. According to Allen et al. (2002), China is a significant counterexample to the findings of the existing literature on law, finance and growth. Despite its underdeveloped legal and financial system, China is one of the largest and fastest growing economies in the world.

Unlike some former communist countries, where the government undertook the massive privatization of SOEs, one striking feature of the Chinese transition is its fast industrial growth without large-scale direct privatization in the reform (Jefferson and Singh, 1999). The Chinese government still maintains ultimate control of the economy and the socialist characteristics of the Chinese economy have been preserved while introducing market-oriented reforms (Garnaut et al., 2005; Naughton, 2007; Prasad, 2004). Decentralization began with the movement of administrative planning to local authorities from the late 1970s (Naughton, 1995, 2007). Since then there has been a deeper movement away from State planning and State-run institutions. This has allowed firms greater autonomy with which to base decisions increasingly

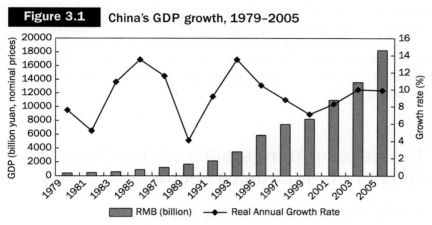

Figure 3.1 China's GDP growth, 1979–2005

Source: National Bureau of Statistics (2006)

on market forces, particularly with regard to resource allocation (Duncan and Huang, 1998).

Therefore, developing competition without large-scale privatization is a typical feature of China's enterprise reform: competition is viewed as a strong support of economic growth, and efficient firms are selected to grow while inefficient firms will inevitably fail (Liu and Garino, 2001a). Market competition eliminates inefficient firms through a 'selection effect', which raises both productive and allocative efficiency by lowering prices and costs (Jefferson and Rawski, 1994).

However, this does not mean that there was no large-scale change in ownership arrangements within firms (Haggard and Huang, 2008; Jefferson and Singh, 1999; Wen, 2002). The government relaxed its monopoly over industry, which was opened to new entrants after 1979. This heralded the rapid growth of the non-State sector, generated a strengthened economy and facilitated reforms of the State sector. Jefferson and Singh (1999) argue that changes in ownership arrangements within firms and the entry of firms with new ownership types have made a large contribution to rapid industrial growth.

The elements involved in the process of the reform were in many ways interrelated. The entry of new firms followed the loosening of the State monopoly, which in turn led to the embracing of market prices and greater competition (Wu, 2005). Increased incentives followed, as did an improvement in State sector performance (Naughton, 2007). However, the non-State sector grew more quickly, as it was more equal to the challenges than the State sector. Ultimately, the non-State sector grew to the point where the State sector became less dominant in the economy (Haggard and Huang, 2008; Naughton, 1995).

The gradual character of Chinese market formation implied that the conditions affecting different industrial sectors varied during the reform period. The following section reviews the development of the Chinese CE sector.

The development stages of the Chinese consumer electronics sector

A brief introduction to the sector

The electronics industry may be subdivided roughly into three categories in the Southeast and East Asian context: consumer electronics, components and industrial products. Consumer electronics may be further classified into matured and fad (fashion) products. The main characteristics of the former (for example, radio broadcast receivers) are their longer life span and relatively stable demand. By contrast, fad products (for example, TV games) have a shorter life cycle with an extremely volatile demand. The definition of the CE changes with the times. According to the president of the Consumer Electronics Association of the US, CE refers to any device containing an electronic circuit board that is intended for everyday use by individuals (Miao, 2007). More information technology and telecommunication manufacturers have joined the CE sector, reducing the dominance of audio and video products in the CE sector (Miao 2007).

The electronics sector occupies an important position in China. First, since electronic products can fulfil a variety of human needs, there is a burgeoning demand for them within Chinese and international markets. Second, the modernization of technology across all segments of this industrial sector is synonymous with the application of electronics systems to industrial processes and capital equipment.

The CE sector in China has been one of the fastest growing industrial sectors in China during the past two or more decades. Over this period, the sector has experienced many changes in production, market composition, firm behaviour, ownership structure and level of government intervention. For this reason, the sector is representative of the manufacturing sector at large in the process of economic transition (Jiang, 2001c).

Four phases of development of the Chinese consumer electronics sector

We carried out the fieldwork on which this book is based between 2006 and 2007. The development and challenges for the CE sector since the late 1970s can be classified into four different phases: from the end of 1970s to the mid-1980s, from the mid-1980s to the early 1990s, from the early 1990s to the end of the 1990s, and from the end of the 1990s to 2007. In each period, State control over the Chinese economy was further relaxed and the economy became increasingly market-oriented (Naughton, 1997). There was a progressive shift from the decentralization of planning control over SOEs to the encouragement of non-State-owned enterprises.

Since the market-oriented reform, the Chinese electronics industry has become a pillar of success and is one of the largest industries in China in terms of sales revenues. It accounts for the largest amount of foreign direct investment (FDI) in China (Pecht and Chan, 2004; Zhang and Parker, 2001). The growth of the electronics industry stems from the rapid growth in CE, which was almost non-existent at the end of the 1970s (Pollack, 1985). By 2002 China ranked third in the production of household electronics appliances, following the United States and Japan (OECD, 2002). The year 2006 witnessed the upgrade of China's CE sector. The pressure of cost control generated by global competition pushed the transfer of the manufacturing of CE products from other countries to China. China has become the most important production base for many CE products in the world (CCIDConsulting, 2007). The Chinese CE sector experienced a rapid growth similar to that of the economy as a whole (Zhang and Parker, 2004). As one of the most successful sectors in China, however, rapid development also brings challenges (Pecht and Chan, 2004).

Phase 1: from the end of 1970s to the mid-1980s

During this period, China had a 'taste' of the market (Wen, 2002). In 1978, Chinese reform commenced with the 'decollectivization' of agriculture and a shift from a planned economy towards a more market-oriented economy (Mina and Perkins, 1997). Poor economic performance and its consequence for the livelihood of Chinese people motivated the Chinese government to initiate the reform (Chow, 2002; Groves et al., 1994; Jefferson and Rawski, 1994). The 'open door' policy was an integral part of Chinese reform. One of the objectives of the Chinese 'open door' policy was the integration of its domestic economy with that of the world (Lardy, 2002; OECD, 2002, 2003). Foreign direct investment (FDI) during this period was welcome but highly regulated (Chai, 1997; Chow, 2002). Joint ventures were preferred over wholly foreign-owned foreign enterprises (Chai, 1997). Industrial reform also began in 1978. The emphasis of industrial reform was on expanding State enterprise autonomy while retaining dominant State ownership (Chai, 1997). Thus reform during this period was characterized by elements of the market economy that grew rapidly, but within an environment where the basic elements of the planned economy remained dominant (Wang, 1994).

In order to improve people's living standards, a readjustment of the industrial structure was required (Naughton, 1997). As one of the measures to address structural problems and the imbalance between consumption and accumulation, China transformed its electronics industry from being mainly military-oriented to a more civilian-oriented industry from the end of 1970s (Naughton, 1995, 1997). Moreover, the 'open door' policy brought imported CE products into China. There was a 'highly favorable response' to these products from consumers after decades of shortages of consumer goods in China (Zhang and Parker, 2001). Supplies, however, were scanty, due to limited production capacities. The discrepancy between demand and supply, along with encouragement from the government, created a sellers' market (Zhang and Parker, 2004), which led to the large-scale entry of new firms into this sector. However, the products of domestic firms were limited to simple goods such as radios and black-and-white TVs (*Chinese Electronics Industry Yearbook*, 1984). The result of a greater demand for the products was that managers focused mainly on quantitative targets (Jefferson and Rawski, 1994).

In 1985, the share of sales from SOEs in the Chinese electronics industry was 74 per cent (*Chinese Electronics Industry Yearbook*, 1986). Including the collective-owned enterprises (the enterprises controlled by the local government), the share of sales from electronics enterprises controlled by the government in 1985 was 92 per cent (*Chinese Electronics Industry Yearbook*, 1986). Although the main emphasis on enterprise reform between 1978 and 1985 was concerned with the redistribution of enterprise incomes and the expansion of operational autonomy in order to improve the performance of the enterprises, these reforms failed to achieve the desired improvements (Wei, 2003). The SOEs were still agents of the State's economic bureaucracy. Output and other goals, rather than profitability, influenced the growth strategies of firms. In order to change the output-oriented strategy, official economic policy switched from quantity-oriented economic growth to quality-oriented growth. SOEs in this industry took advantage of a loose credit policy to import foreign technology (Huchet, 1997). At this time, however, the focus was mostly on hardware technologies. Soft technologies such as management know-how were neglected. Coupled with a highly bureaucratic environment and provincially compartmentalized industrial system,[1] the neglect of new managerial technologies resulted in poor effectiveness for most of the foreign equipment purchases (Huchet, 1997). As a result, production of the CE products increased without extensive technological upgrading (Simon, 1992).

Phase 2: from the mid-1980s to the early 1990s

From 1985, the emphasis in economic reform in China shifted from agriculture to the manufacturing sectors (Jefferson and Rawski, 1994; Jefferson et al., 1996; Naughton, 1995, 2007). This was an important period of ideological shift towards a market system (Wen, 2002). From this time, Chinese reform was increasingly in favour of the market. The success of agricultural reform emancipated ideas on economic reform and the Chinese government tried to remake economic institutions (Wang, 1994). Stock exchanges in Shanghai and Shenzhen were established in 1990 and 1991 respectively. Restrictions on FDI were further relaxed and the setting up of wholly foreign-owned enterprises in China became easier after the mid-1980s (Chai, 1997; Chow, 2002). The realized amount of FDI inflow in 1989 was US$3393 million compared to US$916 million in 1983 (OECD, 2003). Further reform of the SOEs was introduced with the 'dual-track' approach and 'contract management responsibility system' (CMRS) after 1985. The dual-track approach was applied to ownership reforms. It split the production of the SOEs into a central plan component and another outside of the plan (Naughton, 1995). On one track, market-oriented institutions emerged in a parallel economy comprising non-State enterprises with diverse forms of ownership. On the other track, the SOEs were retained. The CMRS represented a formal contract between the enterprise and the State. The CMRS granted the SOE managers more autonomy based on the contracts signed with government agencies (Child, 1990; Child and Lu, 1990; Warner, 2002). Shareholding reforms also became more widespread (Green, 2003).

At this stage management and corporate governance initiatives introduced various trials of a management responsibility system and corporatization[2] (*gongsihua*). Although the CMRC achieved some success increasing the autonomy of SOE managers, poor performance was not penalized (Tenev et al., 2002). Trials of corporatization from the mid-1980s aimed to increase the autonomy of the SOEs while maintaining the 'oversight' of the State (OECD, 2002; Wei, 2003). China borrowed Western corporate concepts and practices. However, there were ideological controversies about corporatization in China, which many in the Party feared would lead to privatization (Wei, 2003).

In the early 1990s, the composition of the sector saw the emergence of new firms (Pecht and Chan, 2004). Because of diminished protection from the government, under-sized and under-capitalized firms were acquired by the more successful firms (Huchet and Richet, 2002). Some of the best known companies in the sector lost their leading positions because of 'serious errors of strategy in terms of diversification' and 'faulty assimilation of foreign technology', as well as the withdrawal of government privileges, such as easy access to low-cost loans from State-owned banks (Huchet and Richet, 2002). In their place, new groups of companies emerged, which became highly efficient and internationally competitive (OECD, 2002). The highly competitive environment forced the newly-emerged companies to manufacture products with higher quality, new styles and lower prices (Shi and Zhao, 2001). These new firms not only improved their product quality, but also reorganized their internal structure, production, marketing and sales systems (Simon, 1992).

Phase 3: from the early 1990s to the end of the 1990s

From the early 1990s, Chinese economic reform brought progress to many parts of the economy (Wang, 1994). New laws were passed, such as the Chinese Company Law in 1994 and the Contract Law in 1999, which improved the legal foundations for business. In 1997, the reform of SOEs through corporatization was given a push at the Fifteenth National Congress. Party leader Jiang Zemin proclaimed: 'The joint stock system is a form of capital organization of modern enterprises, which is favourable to separating ownership from management and raising the efficiency of the operation of enterprise and capital. It can be used both under capitalism and under socialism' (Jiang, 1997). The importance of reforming the SOEs was highlighted. During this period, the SOEs' share of industrial output continued to decline, down to 28 per cent in 1998 compared to 64.9 per cent in 1985 (*Chinese Statistics Yearbook*, 1999). After 1992, FDI in China increased dramatically (Lardy, 1996, 2002). China became the largest recipient of FDI among all developing countries from 1993 (OECD, 2003). In the 1990s, the proportion of wholly foreign-owned enterprises and the number of large, relatively high-technology projects initiated by multinational companies grew steadily (OECD, 2003).

This stage saw a marked change in further ownership diversification (Tenev et al., 2002). According to Huchet and Richet (2002), privatization, mergers and bankruptcy constituted the major axis of the restructuring of small SOEs in this sector because they suffered from a system of corporate governance that was especially inefficient. For large and leading corporatized SOEs in the CE sector, they had a special position in corporate governance. Owing to their market success, local government agencies were no longer able to exercise leverage over them and the management of these companies had also gained greater independence (Huchet and Richet, 2002). However, the new market concepts had to be adapted to existing institutional constraints (Tenev et al., 2002). For instance, although the Chinese Company Law requires all listed companies to adopt a two-tier board structure comprising a board of directors and a supervisory board, these coexisted with the 'three old committees', the Party Committee, the Workers' Council and the trade union (Tam, 1999).

During this period, the Chinese CE sector continued its rapid development. The total production of colour televisions, radios and audio-cassette recorders ranked first in the world as China built up its mass production capabilities and improved the global competitiveness of its products (Pecht and Chan, 2004). Chinese products also became highly competitive on the home market. The reform measures gave enterprises and local governments the incentives and the ability to seek profit. The entry of many enterprises into the market increased competition, which in turn made it possible to eliminate the chronically short supply of CE products, thus making a buyers' market where competition is the rule (Jiang, 2001a). Rapid development from the 1980s meant that the sector faced overcapacity in the early 1990s (Simon, 1992, 1996). For example, the average urban Chinese household had 1.2 colour TV sets by the end of the 1990s, yet at the end of the 1980s only around half of all urban households had a colour set (*Economist*, 2001). These changes forced firms to respond better to the market.

In the domestic market, the firms lowered the price of the products while upgrading their technology in order to win a greater proportion of the market share. At the same time, they broke into the world market through technology imports, joint ventures and an original equipment manufacturing (OEM) arrangement (Shi and Zhao, 2001). Some successful firms not only entered the markets of Europe, the USA and Japan, but also established a production base and sales network in overseas markets.

Phase 4: from the end of the 1990s to the present

China's entry into the World Trade Organization (WTO) in 2001 opened China's door even more widely and increased pressures for greater economic liberalization. In 2002, China ranked first in attracting FDI (OECD, 2003). At this stage, driven by the need to be competitive in the global economy, China strove to improve its institutional infrastructure. Measures included policy protecting private ownership, a new definition of China's market economy, and the scraping of the rules and regulations that did not conform with WTO norms (OECD, 2002, 2003).

During this period the CE sector experienced increased price liberalization, a lowering of entry barriers for producers and the development of unbridled competition. China has become the world's foremost manufacturer of many CE products such as colour TV sets, colour tubes and cellular phones (Pecht and Chan, 2004). This sector became one of the largest recipients of FDI as more foreign firms established their production bases in China (OECD, 2002). With the deepening of Chinese economic reform, firms in this sector have relied mainly on the market for inputs and the sale of outputs instead of mandatory planning as at the start of economic reform (Zhang and Parker, 2004).

In China, there was a list of priorities for project choices at this stage (see Table 3.1). Since the CE sector tallied with the first four motives

Table 3.1	Priorities of government motives for their choice of industrial projects	

Motives for choosing projects	Occurrence frequency (%)	
	Municipal government	Provincial government
Increasing local revenue	89.3	87.9
Meeting local demand	87.3	85.4
Speeding up economic growth	85.7	84.4
Taking lead among counterparts	83.4	76.9
Achievement in urban development	78.6	75.8
Infrastructure and environment	50.5	54.6

Source: Jiang (2001b :177)

of the government, it was encouraged to speed up development (Jiang, 2001b).

At this stage, more foreign higher-technology CE products had entered the Chinese market. In 2001, the share of industrial value added from joint ventures and wholly foreign-owned enterprises was 54.5 per cent in the electronics industry (*Chinese Electronics Industry Yearbook*, 2002). These foreign firms rely on their technology advantage to compete with domestic firms. Some of them have established not only production bases in China, but also technological centres (OECD, 2002). Chinese firms put greater effort into upgrading their products, improving technologies and reforming their corporate governance due to intense competition from both domestic and foreign competitors (Liu and Woo, 2001; OECD, 2002; Pecht and Chan, 2004). Chinese firms are more aware of increasing their market share through modernizing their equipment and management practices rather than lowering their price (Huchet and Richet, 2002; Tan and Tan, 2003).

Some SOEs, especially large ones, are still an important group of firms in the CE sector in China. These competitive SOEs have not been deliberately discriminated against in the industrial policy of the central or local governments. They have not been supported, either financially or logistically, to the same degree as the major SOEs that dominated their sector during the 1980s (Huchet and Richet, 2002). The management of these companies has also gained relative independence from their parent authority.

Corporate governance has received particular attention since the late 1990s in China (Wei, 2003). Market competition has driven firms to dilute the ownership control of the State in exchange for public funds to finance their growth, which has resulted in an ownership evolution (Liu and Woo, 2001). The shares of gross value of the industrial output of limited liability companies in the Chinese electronics industry were 18 per cent in 2007 compared to 7.09 per cent in 1995 (*Chinese Electronics Industry Yearbooks*, various years). On the one hand, strong competition in their market improved the system of corporate governance of such firms (Huchet and Richet, 2002). On the other hand, these enterprises have much more complex internal structures and are facing external institutions such as the remnants of the old command economy. Reforming all of these institutions is a complex task and far from complete in China (DFAT, 2002).

In summary, the Chinese CE sector has undergone great changes since the end of the 1970s and has emerged as one of the key manufacturing sectors in China's national economy. However, business strategies and corporate governance at firm level in the CE sector and the relationship between business strategies and corporate governance have not been explored. This study will address these questions in detail.

The classification of firms by ownership in China

One of the distinctive features of the Chinese market environment is the competition between firms of very different ownership forms. However, strategic management research has not paid much attention to the ownership issue (Peng et al., 2004), largely because of its origin in research emanating from developed economies. Industry in developed countries is dominated by private ownership. There is therefore far less research on strategic management issues for firms that are State-owned.

In contrast to market economies, the Chinese economic system is characterized by diverse forms of ownership among its enterprises. However, this diversity only emerged over the course of economic reform. In 1980, the Chinese industrial sector consisted almost exclusively of State and collectively-owned enterprises. In 1993, the Chinese government initiated the shareholding programme, which became the principal vehicle for implementing ownership reform of the SOEs (Jefferson and Su, 2006).

Since 2001, the National Statistics Bureau (NSB) has made a further ownership-based distinction among enterprises: State-owned enterprises and enterprises with controlling shares held by the State, collectively owned enterprises, individually owned enterprises, and enterprises of other economic types (see Figure 3.2). The last category includes private companies, shareholding companies and foreign-funded enterprises involving Hong Kong, Macao and Taiwan investments (Bian and Zhang, 2006).

However, whether Chinese statistics are capable of providing a reliable account of the categorization of firms is an open question (Perotti et al., 1999). The issue of shareholding companies in China is complicated. The shareholding firms typically fall into State-dominated, collectively dominated and privately dominated shareholding companies. Therefore, treating the shareholding companies as a single kind of ownership based on the official categorization suffers biases, which underestimate the likely impact of quite diverse ownership traits on the approach to doing business and organizing the firm.

In addition, this study does not use the shareholder information as the main criteria for categorizing the firms as some research does (Dougherty

Figure 3.2 Categories of firms in China

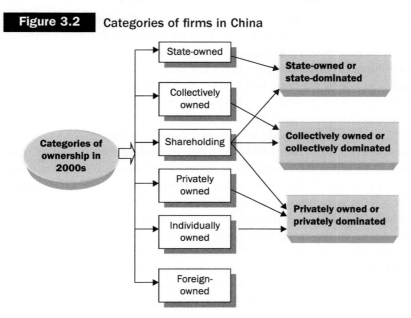

Source: Based on NSB classification (2001).

Note: The foreign-funded enterprises are not included in the graph since they are not the focus of study here.

et al., 2007). According to the controlling shareholders in shareholding companies, firms can be separated, whether by the State (directly or indirectly through legal persons), a collective (local government), or a private entity (individuals, domestic legal persons, or foreign companies) that controls the firm. However, it is hard to decide the ownership of firms based on this criterion in China. With the further development of China's reform, central government has delegated its control to local government. In most cases, the ultimate owner of the listed firms of the SOEs is local government. In addition, the definition of legal person (LP) shares is not clear. According to Liu and Sun (2003), although LP shares are those owned by domestic institutions, the official classification fails to identify whether the ownership status of the LP shares is State-owned or non-State-owned institutions. The owners of the LP shares might well be institutions controlled by central or local government. Share information alone is therefore inadequate to decide the ownership type of firms in China.

Rather than using the official firm registration status that is commonly shown in Chinese statistical publications or share information to look at ownership, this study classifies firms by nature of the ultimate controlling shareholder. For example, many SOEs have been transformed, partly or fully, into shareholding companies classified into the category of 'other ownership', although the State holds the majority of shares. These types of firms in which the State is the de facto majority investor and ultimate owner should be counted as a part of the SOE sector. There are many overlapping characteristics between listed SOEs and traditional SOEs. In this case, this study regards State-controlled shareholding companies as State-owned. SOEs come in two categories: absolute control and relative control. Absolute State control implies that the State accounts for more than 50 per cent of the total capital. Relative State control implies that although the state holds less than 50 per cent of total capital, its share is relatively large compared to the shares of other ownership categories that hold shares in the enterprise (Holz and Lin, 2001). The same pattern holds for the categorization of COEs and DPOEs. Since most of the listed companies in China are only just one part of the parent company, which is the largest shareholder of the listed companies, this study's judgement of the nature of the shareholding companies is based on the nature of their intermediate shareholder, which can be obtained from the registration information of NSB. According to this analysis, Chinese indigenous firms are divided into three categories: State-owned (dominated), collectively owned (dominated) and privately owned (dominated) firms (see Figure 3.2).

The development of State-owned enterprises (SOEs)

The importance of SOEs in the Chinese economy in the past resided in their status as a significant provider of government fiscal revenue, employment and social services. Although the SOEs' share of gross industrial output value declined throughout the reform period (recently less than 30 per cent of the overall production in China compared with 77.6 per cent in 1978) (Hassard et al., 2007), many SOEs were able to compete in a more open market-oriented economy. These companies were set to play a significant role in the economy for many years to come.

As an organizational form, the SOEs are unique in the role played by the State (Peng et al., 2004). First, the owner of an SOE is the State, or as typically phrased in China, the 'people as a whole' (Wei, 2003). The owner of SOEs is not a person or even a single organization

(Aharoni, 1981). The bureaucrats represent the State or the 'people as a whole' (*quanmin*) as the owner of the enterprise. The de facto absence of identified owners and the lack of clarity in property rights attached to a firm increase managerial discretion in a potentially adverse way because the monitoring is much more difficult in SOEs. Owing to the high agency costs, managers are able to engage in opportunistic behaviour guided by self-interest rather than the benefit of the firm (Tang and Ward, 2003).

Second, the SOEs have had multiple objectives in the past, including the welfare of the employees, running their own shops, kindergartens and schools as well as maintaining a cradle-to-grave welfare schemes for their employees (Vashist, 2004). Managers had little authority over research and development, product innovation, investment planning, marketing or even such routines as production scheduling, material purchases, wage structures and employment levels (Tan and Tan, 2003). The evaluative process of enterprise performance emphasizes meeting production targets set down by higher authorities. Profitability influenced neither the incomes of executives and workers nor the growth prospects of firms. The lack of autonomy over managing firms discouraged managers from adopting innovative measures to improve their firms' performance.

Third, the government at one time heavily subsidized the SOEs in order to allow them to achieve some outcomes, such as welfare goals (Estrin, 2002). Due to their economic and strategic significance, most SOEs were not allowed to fail (Tan and Tan, 2003). SOEs received privileged access to key resources, especially subsidized credit and government financial transfers to support loss-making SOEs, which increased inefficient resource allocation. This kind of subsidy constrained the economy's ability to generate balanced growth (World Bank Group, 2001).

However, market-oriented reform in China has reshaped the competitive landscape for SOEs (Tan and Tan, 2003). Under this reform, the relationship between the government and the firms started to change. The Chinese government granted SOEs unprecedented autonomy as well as the financial independence to compete (Jefferson and Rawski, 1994). Although the State still owns a majority of the stakes, many SOEs have been given some market or market-like incentives (World Bank Group, 2001). The profit orientation has gradually taken root in many SOEs (Yang and Zhang, 2003). The SOEs increasingly make decisions according to supply and demand, rather than adhering to State administrative orders (Peng, 2004). Such newly acquired autonomy and flexibility have motivated SOEs to build the resources and capabilities to compete.

The development of collectively owned enterprises (COEs)

There are many uncertainties in the published definition of COEs (Ding, 1998). Definitions of COEs in administrative regulations do not always correspond. Analysed from the angle of enterprise ownership, COEs in the early 1980s were defined as owned by 'members of a locality such as a city, a commune, or even a small group of people such as a brigade or an urban neighborhood who participated in the enterprise as shareholders or workers' (Sit, 1983: 86). Based on this form of ownership structure, township and village enterprises (TVEs), together with joint urban enterprises and joint rural enterprises, constitute COEs (Luo et al., 1998).

As a firm type, COEs in China have received less attention from scholars than have SOEs and privately owned enterprises. Although there is literature on COEs (Pan and Park, 1998), more attention is paid to the TVEs (Luo et al., 1998; Morgan, 1994). The TVEs are either collectively established by or initially based on and closely associated with rural communities (townships and villages). In fact, many TVEs were privately owned firms in disguise (Dickson, 2003; Haggard and Huang, 2008; Sun, 2002b). These privately owned firms were registered in the name of COEs in order to gain some political protection. This phenomenon is called wearing a 'red cap' (Dickson, 2003; World Bank, 2000). By establishing a partnership with local governments, TVE entrepreneurs essentially use the organizational form as a 'boundary blurring' strategy in order to seek institutional protection. In this case, privately owned firms use the category of the COE as a shield to protect themselves from environmental uncertainties (Peng, 1997).

The large-sized urban COEs are different from TVEs although they are within the same ownership category. Large urban collectives are much closer to the 'formal' sector and may often not differ much in their organization from formal SOEs (Shi, 1998; Sit 1983; Walder, 1994). The government treats them more or less as State enterprises due to their unclear ownership. According to Shi (1998), these large urban COEs have been merged and dismantled several times and have undertaken several investment projects financed by local governments. As a result, although these COEs still remain collectively owned, they are in practice owned by the local State. 'As far as property rights assignments go, the common administrative distinction between State and collective ownership is virtually without meaning' (Walder, 1994). When a firm constitutes the primary source of revenues for a local government, it receives enormous aid from that local government including financing, access to resources, risk diversification, and the like (Li and Zhou, 2005). Since large collectives are still playing a significant role in providing employment, supplying products and contributing tax revenues to the local government, more assistance from the local government helps to create a more favourable task environment for large collectives. The local government is willing to provide some support to large COEs for developing state-of-the-art technologies – typically in the form of grants (conditioned on research results) and loans, both low-interest and interest-free. Their characteristics are of the government-oriented firm type and they constitute a component of the public sector.

With deepening economic reform in China, the number of COEs has declined. The number of large and medium-sized urban COEs fell from 4068 in 1994 to 3408 in 1999. The 16 per cent decline in the number of collective enterprises is striking (Jefferson et al., 2003). The decline in the COE sector is also evident in the decreasing number of employees working in COEs between 1978 and 2007 (see Figure 3.3).

The reason for the decline lies in the policy of encouragement of private ownership. Under the planned economy, privately owned and foreign owned enterprises were severely restricted. COEs were generally viewed as an immature form of State ownership falling between private and State ownership of the means of production (Tang and Ma, 1985). COEs grew out of the handicraft sector which had been a supplementary source of employment in the city and the countryside (Sit, 1983). The COEs were

Figure 3.3 Employees by enterprise ownership (%), 1978–2007

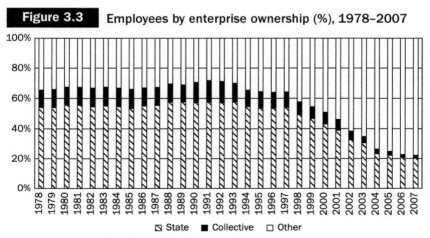

⊠ State   ■ Collective   □ Other

Source: Chinese Statistics Yearbook (various years).

supposed ultimately to be replaced by SOEs. With the original aim of the role and nature of the COEs phased out, most of these COEs have been transformed into privately owned enterprises with the encouragement of the municipal government (Gregory et al., 2000). Many of the 'red hat' pseudo-collective firms started to take off the 'red hat' in the late 1990s, which accounted for some of the decline in the number of COEs and the rise in DPOEs (Dickson, 2003, 2007).

The development of domestic privately owned enterprises (DPOEs)

Over the past three decades, China has made major reforms in the course of its transition from a command economy to a more market-driven economy, which in the last decade or more has seen significantly increased private ownership of enterprises. Fan (2000) holds that the most important issue of Chinese transition is the development of the non-State sector, rather than the reform of the State sector. The non-State sector has been the major contributor to the market-oriented transition and overall economic growth for the past 20 years. The non-State sector has also pushed the reform of State-owned enterprises, challenging their monopolization of some sectors and increasing competition, creating employment and fuelling the emergence of entrepreneurs. Thus, to invigorate enterprises, State ownership has needed to be divested (Fan, 2000). Although early reforms gave the non-public economy a fresh start, privately owned enterprises did not become legal until the Chinese constitution was amended in the late 1990s.

As discussed in the previous section, many DPOEs in the past would rather be registered as COEs or TVEs than be under private control (Dickson, 2003). Collective or TVE status alleviated policy constraints associated with private status. These so-called red-hat firms enjoyed political protection and easier access to the resources like land, assets, finance and markets that were exclusively reserved for SOEs (Haggard and Huang, 2008). In the late 1990s, amendments to China's constitution declared that the private sector is an 'essential part' of the socialist market economy. Privately owned enterprises thereafter became a major driving force in the economy, and private entrepreneurs as a new social class began to attract enormous interest across Chinese society (Dickson, 2003). With the improvement of the macro environment and encouragement of the government, the private economy in China has shown rapid development (see Figure 3.4).

| Figure 3.4 | Number of registered private firms in China, 1989–2007 |

Number (1000)

Source: *Chinese Statistics Yearbook* (various years).

The factors accounting for DPOEs' growth and development include the following: (a) Hard budget constraints on DPOEs. Domestic privately owned enterprises generally face hard budget constraints, that is, they need to be responsible for their own financial performance and debts (Naughton, 2007). In order to ensure their survival in the market place, they exhibit a strong propensity for risk-taking, innovation and proactiveness in their investment decisions (Tan, 2002); (b) Flexibility due to small size. They are small but nimble and good at market orientation in order to grasp business opportunities emerging in all the markets of products, labour and capital, both domestic and international (Peng et al., 2004; Sun, 2002a, 2002b). The DPOEs are self-motivated and respond flexibly to changing policies and regulations while diversifying risk and avoiding excessive taxation and competition, which may allow them to outmanoeuvre more established firms such as SOEs (Peng et al., 2004).

Despite DPOEs comprising the most rapidly growing sector of China's economy, many such firms remain subject to a variety of policy and economic constraints (Bian and Zhang, 2006; Garnaut et al., 2001; Gregory et al., 2000; Haggard and Huang, 2008). Due to the influence of the planned economy, the status of most DPOEs is not high and they rank low in the government's priority list (Tan, 2002). Finance for small and medium-sized private enterprises is insufficient. DPOEs lack access to land and the bank credit accorded to the SOEs. They also lack the protection afforded by laws. The restrictive stance toward the DPOEs prevents them from attracting the finance and skills they need to grow (Garnaut et al., 2005). Not having State support, the DPOEs operate at a disadvantage in coping with the regulatory framework and the financial system. For example, they have less support from the government, poor investment in research and development (R&D), and limited access to critical resources. Yet, DPOEs in China operate in industries that happen to be expanding robustly in the current stage of China's development, calling for lower investment and technology (Jiang, 2001c).

The shareholding programme in China

Shares of ownership of listed firms in China

Shareholding stock companies and stock markets did not exist until the early 1990s, when the Chinese government decided to restructure further the SOE industrial sector. Shanghai and Shenzhen Stock Exchanges were opened in 1990 and 1991 respectively. Since then, they have steadily grown from a handful of listed firms to 1576 firms in December 2007 (see Figure 3.5).

In the 1990s, China's central government had initially promoted shareholding cooperatives, which represented a temporary compromise among local governments, enterprise managers and individual entrepreneurs. In 1993, the Chinese government initiated a shareholding

| Figure 3.5 | Number of firms listed on the Shanghai and Shenzhen stock exchanges, 1990–2007 |

Source: Chinese Statistics Yearbook (various years).

programme. The shareholding system (*gufen tixi*) is a modern approach to capital allocation that has the potential to improve the productivity of capital and enterprises.

Since 1994, the Chinese government has sought to transform SOEs into 'modern enterprises' with 'clarified property rights, clearly defined responsibility and authority; separation of enterprise from the government; and scientific internal management'. The policy shift to promote de facto privatization was endorsed in 1997 at the Fifteenth Congress of the CCP, where Jiang Zemin told delegates that the stock-holding form of listed enterprise was also appropriate for a 'socialist market' economy.

A mixed form of ownership is characteristic of a typical listed shareholding company in China. Shares are classified as A-shares designated for domestic investors and selected foreign institutional investors, and B-, H-, and N-shares designated for overseas investors. A-shares are further divided into State shares, legal person shares, tradable A-shares and employee shares.

State shares are those held by the State including the central government and local governments. State shares can also be held by the parent of a listed company, typically an SOE (Tenev et al., 2002). State shares are not permitted for trading at the two exchanges, but transferable to domestic institutions, upon approval of the China Securities Regulatory Commission (CSRC)[3] (Xu and Wang, 1999). In many of the listed firms, the State is the largest or majority shareholder.

Legal person shares are those held by domestic legal entities and institutions such as other stock companies, State-owned enterprises, and non-bank financial institutions. 'State-owned legal person shares' is a sub-category of legal person shares. It refers to shares held by institutions in which the State is the majority owner but has less than 100 per cent shareholding. Like State shares, legal person shares are not tradable on stock exchanges, but they can be sold to other legal persons. Sales of legal person shares to foreign investors were allowed until suspended in May 1996 (Lin, 2001).

Tradable A-shares are owned and traded mostly by individuals and by some domestic institutions. There is no restriction on the number of shares traded, or on holding periods (Chinese Company Law, 2005). It is required, however, that tradable A-shares should account for no less than 25 per cent of total outstanding shares when a company makes its Initial Public Offering (IPO)[4] (Chinese Company Law, 2005). These shares are the only type of equity that may be traded among domestic investors on the two exchanges. The often cited 'public shares' are shares held by individual investors that are tradable.

B-, H-and N-shares at first would only be held and traded by foreign investors, but since 2001 Chinese investors can also hold. The market for B-shares is separated from the A-share market. The price of these shares is quoted mainly in US dollars on the Shanghai Stock Exchange and in Hong Kong dollars on the Shenzhen Stock Exchange (Tenev at al., 2002). H- and N-shares are similar to B-shares in nature, except that they are issued and traded on the Hong Kong Stock Exchange and New York Stock Exchange respectively.

Employee shares are a unique feature of the Chinese shareholding system and different from an employee stock ownership in the USA. They represent accumulated profits retained by the pre-IPO entity under the Contract Responsibility System and are collectively owned by the employees of the company (Xu and Wang, 1997, 1999). They are not tradable at the time of listing and are managed by either an investment management committee or a staff union. Employee shares are registered under the title of the labour union of the company, which also represents shareholding employees to exercise their rights. After a holding period of 6 to 12 months, the company may file with the CSRC, allowing its employees to sell the shares in the open market (Tenev et al., 2002). Most listed companies do not have employee shares and where such shares do exist they typically account for a very small fraction of the total shares outstanding.

China's biggest challenge with the shareholding programme is coping with the fact that the State is the controlling shareholder in the vast majority of listed companies (Weinstein, 2008). Despite their public listing, many large SOEs remained overwhelmingly State-controlled (Green, 2003; Lu, 2006b; Naughton, 2007). In China, although tradable shares are gradually increasing, direct and indirect State ownership in aggregate has reduced very slowly (Hovey, 2004). Most listed firms in China are still State-dominated and controlled. In the Shanghai Stock Exchange, the State directly held a 50.29 per cent stake in listed firms, in addition to holding the largest proportion of the LP shares in May 2007. Thus more than one-half of the shares have been excluded from the capital market and are not permitted to be traded (Hu and Goergen, 2001; Lu, 2006a, 2006b; Tam, 1999).

There are documented abuses by the State as the largest shareholder including taking out soft loans from listed firms on a long-term basis, using listed firms as guarantors for bank loans, and selling assets at unfair prices, without an appraisal by an independent evaluator (Feinerman, 2007). The following statute and regulations attempt to address these abuses.

Relevant statutes and regulations in China

Since the establishment of China's securities regulator, the China Securities Regulatory Commission (CSRC), in 1992, passed more than 300 laws and directives concerning the securities and futures market have been issued. The key legal framework for corporate governance in China consists of (1) Chinese Company Law, issued in December 1994; (2) Establishment of Independent Directors Systems by Listed Companies Guiding Opinion, published by CSRC in August 2001; (3) Code of Corporate Governance for Listed Companies, released in January 2002 by the CSRC and the State Economic and Trade Commission; and (4) Administrative Measures on the Split Share Structure Reform of Listed Companies, announced in 2005. The main roles and functions of these laws and policies are as follows.

The Company Law was the first attempt since 1949 to create limited liability companies without regard to the nature of ownership (Feinerman, 2007). Under the Chinese Company Law the traditional SOEs would be corporatized either into a joint stock corporation or a limited liability company (Clarke, 2003; Naughton, 2007). Corporatization under the Company Law aligned the interests of the managers with their government owners. The executive directors of State-controlled listed firms were appointed by the government in its capacity as a shareholder. The board of directors then gave direction and oversight, and hired and fired managers. In this way, the State was to exercise its ownership rights by redefining its role as a pure shareholder (Naughton, 2007). However, this legal framework – even after the recent amendment of the Company Law – remains weak in addressing some important issues. For example, under the amended Company Law in 2005, the supervisory board still does not have any substantial power. Articles 52 and 118 require that at least one-third of the supervisory board members must be elected by employees (Chinese Company Law, 2005). In practice, the salaries and promotion of the employee supervisory board members are determined by the directors and managers (Feinerman, 2007). Employee supervisory board members, in turn, typically remain loyal to the interests of the directors and managers. The supervisory boards often fail to exercise effective supervision under the existing equity structure dominated by parent SOEs or government agencies in charge (Tenev et al., 2002).

As corporate governance reform progressed, some scholars argued that an independent director institution would solve the numerous corporate governance problems entangling Chinese listed firms (Weinstein, 2008). The independent director system was initially adopted by a small number of Chinese companies that were listed in the Hong Kong stock market (the H-share companies) after 1993. The concept of independent directors officially appeared in China as early as December 1997, when the *Guidelines for the Incorporation of Listed Companies* allowed Chinese companies to appoint independent directors on a voluntary basis. State control of most listed firms severely curtailed any practical role for independent directors if firms were inclined to appoint them, and most were not.

The use of independent directors was effectively optional (Weinstein, 2008). Only a very small number of Chinese listed companies had appointed independent directors before 2001 because of the absence of any mandatory requirement (Fan, 2000; Li, 2001). On 6 August 2001, the CSRC issued The Establishment of Independent Directors Systems by Listed Companies Guiding Opinion (hereinafter the 'Guiding Opinion'). This landmark document mandated that by 30 June 2003, the board would comprise at least one-third independent directors, one of whom was required to be an accounting or financial professional. It also included specific rules on the definition of independent directors as well as a clarification of the powers of independent directors (China Corporate Governance Survey, 2007). Even though the Guiding Opinion appears to grant independent directors power, the controlling shareholders – often the State in State-controlled firms – appoint their acquaintances to serve as independent directors. This practice hardly satisfies the goals of an independent director system (Miles, 2006; Tenev et al., 2002). A 1999 survey showed that only about 3 per cent of all directors had some degree of independence; in 2003, following issuance of the Guiding Opinion, the average company still had only three independent directors (Lu, 2005). China is among the few nations that have established both the independent director and supervisory board institutions. It is more cost-effective to choose between the two institutions because their oversight functions overlap.

On 7 January 2002, the CSRC State Economic and Trade Commission reaffirmed the independent director system when it promulgated the Code of Corporate Governance for Chinese Listed Companies (hereinafter the Code). Article 49 requires listed companies to introduce independent directors who do not hold any other positions within the company (Code of Corporate Governance for Listed Companies (CSRC), 2002). While the Code and the Company Law formally institutionalize the requirement for independent directors, the Guiding Opinion (2001) is the most important document regulating the independent director system in China (Lu, 2006a; Yuan, 2007). The Code also attempts to strengthen the roles of the supervisors. Articles 60 and 61 state that members of the board of supervisors must be permitted access to information related to operational status and be allowed to hire independent intermediary agencies for professional consultation, without interference from other company employees (CSRC, 2002). While there is some pressure on companies to follow another agency theory prescription to abandon CEO duality – the practice of one person serving both as a firm's CEO and board chair (Song et al., 2006) – there is no mandatory requirement in the Code to separate the roles of CEO and chairman of the board. In other words, listed firms in China have autonomy in either combining or splitting the two top positions (Peng et al., 2007). In reality, the percentage of listed firms practising CEO duality has been decreasing, from about 60 per cent in the early 1990s to about 20.9 per cent in 2006 (*www.news.xinhuanet.com*, accessed on 26 September 2006). However, some argue that having a separate chairman and CEO under these circumstances may be of little value (China Corporate Governance Survey, 2007). Under the Chinese model, most Chinese companies have a single dominant shareholder – for example, the State – that generally appoints both the chairman and CEO positions. Even if a separation were mandated and enforced, it is questionable whether enforcement would be effective.

The new initiative (Administrative Measures on the Split Share Structure Reform of Listed Companies) provides the ground rules to allow for the conversion of shares. The first area for reform in China's corporate governance is the highly concentrated ownership structure in Chinese companies (Tam, 2002; Tenev et al., 2002; Wei, 2003). In fact, the official term used in China is 'share ownership scheme' rather than 'privatization' (Sun et al., 2002). The three largest shareholders accounted for 60 to 80 per cent of total shares in half of all the listed Chinese firms (Feinerman, 2007). Put into effect by mid-September 2005, the share reform concerning the relaxation of the ownership constraints allows the sale of State-owned shares and attempts to address the concentrated ownership issue. According to a report of the 'Notice Relevant to Pilot Reform of the Segmented Share Structure of Listed Companies' of 29 April 2005, China wants to encourage firms to convert their non-tradable State shares into tradable A-shares. The share reform measures facilitate the government to improve the transparency of the stock market and to stop insider trading.

China's corporate governance reform has flaws but is a positive step to better governance. The issuance of these regulations and the adoption of the legal framework for corporate governance have led the way for board directors and corporate managers to adopt sound corporate governance practices. For the vast majority that has State controlling interests, China needs to take another step towards standardization, enforcement and expansion of its corporate governance laws and regulations (Weinstein, 2008).

To sum up, the shareholding system reform in China has resulted in a shift away from central planning to a market orientation. Although there are many problems that must be resolved in order to become effective free market corporations, a competitive market characterized by the predominance of public ownership represents an important aspect of a modern 'socialist market economy' with 'Chinese characteristics' (Liu and Garino, 2001a).

Background to the case firms

Sampling of the cases

The choice of cases for study is critical, requiring great attention to the appropriate procedure for choosing them (Lieberson, 1992; Yin, 1984, 2003). A key feature of this research design is the use of the multiple cases and data analysis, based on comparing the findings of the individual cases. The sampling of the cases in the study was purposive, rather than random. To expose the theoretical speculations to a new empirical phenomenon, eight companies were analysed in depth. Eight companies do not provide a large sample. However, researchers do not need a large number of cases for an in-depth study (Gummesson, 2000).

Selecting three categories of firms

This study focused on Chinese indigenous firms, namely SOEs, COEs and DPOEs. These enterprises have not been subject to extensive field research (Peng, 2005; Tsui, 2004). Although the growth of foreign-invested firms in China (for example, joint ventures and wholly owned subsidiaries) has also been impressive during the same period, these have been analysed extensively.

Based on the official categorization of firms discussed earlier in Chapter 3, the categories of firms in China include State-owned enterprises, collectively owned enterprises, individually owned enterprises, joint-ownership companies, private companies, shareholding companies, and foreign-funded enterprises. Chinese indigenous firms in this study are divided into three categories: State-owned or State-dominated firms, collectively owned or collectively dominated firms and privately owned or privately dominated firms.

As discussed previously, there are overlapping characteristics between the firms. For example, many of the SOEs have been transformed, partly or fully, into shareholding companies classified into the category of 'other ownership', despite the State holding the majority of shares. Actually, the firms in which the State is often the de facto majority investor or controller should be counted as a part of the SOE sector (Perotti et al., 1999). Therefore, the official categorization suffers bias.

Criteria for selection of cases

The selection of the eight case study firms was based on their broad or dominant ownership type, location and other criteria.

First, the cases were selected to be representative, comparable and manageable. By the term 'representative', this study refers to different ownership forms and technology. Different ownership forms imply variety in their level of government jurisdiction, administrative jurisdictions, and experiences during the reform. The selected firms demonstrate variety in technology, with different growth rates and competitive position. The extent of technology-intensiveness had to be taken into account. The cases selected for this study not only included manufacturers of finished products, but also those providing components for the former. Among the eight cases, some are relatively high-tech, and some are relatively low-tech, but all were selected from within one industry in order to minimize extraneous industry variation that might be derived from differences between sectors, such as government policy and degrees of openness. The term 'comparable' means the intensity of the market for corporate control, and the competitiveness of the product market common to most firms in one industry. Finally, the cases should be manageable, which implies that the selected cases were cooperative and offered easy access to information.

Second, given the size of China, the target firms were located in six cities of three different regions: Qingdao in Shandong province, Nanjing and Taizhou in Jiangsu province, and Hangzhou, Haining and Shangyu in Zhejiang province (see Map 3.1).

Case selection in different areas minimizes area-specific limitation. Jiangsu and Zhejiang dominated the CE sector in terms of output in the early 1980s, while in more recent times Shandong has had the biggest consumer electronic household appliance production base in terms of sales. The DPOEs were chosen from Zhejiang province based on its advanced private sector development. The province has a history of

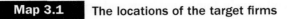

| Map 3.1 | The locations of the target firms |

Note: The dots refer to the cities where the fieldwork was undertaken. 1 is Haining, 2 is Shangyu and 3 is Hangzhou.

pioneering new forms of enterprises. It has been a leader in the development of the private sector in China since its reform. The private sector is most developed in Zhejiang province (Wang, 2006), with its economic output value making up 95 per cent of the province's total (Statistical Bureau of Zhejiang Province, 2004).

Only two COEs instead of three have been included because of the difficulty of finding 'true' COEs due to their blurred definition during the economic transformation. This study's COEs were urban collectives that have received less attention in scholarship. The final sample is indicated in Table 3.2.

Profile of the cases

Case 1: Hisense Group

The Hisense Group (hereafter referred to as Hisense) is a State-owned company and dominant player in the Chinese market for CE products.

Table 3.2	Brief introduction to the eight firms, 2006

Firms in the Chinese consumer electronics sector					
Name	Firm type	Location	Date of establishment	Employment	Main business
Hisense	SOE	Qingdao, Shandong	1969	7500	TV sets
Panda	SOE	Nanjing, Jiangsu	1936	6000	TV sets
Huadong	SOE	Nanjing, Jiangsu	1937	4929	Lighting lamps
Haier	COE	Qingdao, Shandong	1984	30571	Refrigerators
Chunlan	COE	Taizhou, Jiangsu	1985	8475	Air-conditioners
Yankon	DPOE	Shangyu, Zhejiang	1975	4267	Electronic energy-saving lamps
Silan	DPOE	Hangzhou, Zhejiang	1997	710	Microelectronics
Tiantong	DPOE	Haining, Zhejing	1984	3124	Soft magnetic products

Sources: *Chinese Electronics Industry Yearbook,* documents on the firms and interviews (March–July, 2006).

Located in Qingdao, an eastern coastal Chinese city, the headquarters of Hisense was established by the Qingdao municipal government in 1969. In its early years, the company was a small radio factory with a dozen employees.

In the 1970s, the factory shifted its focus to the production of black-and-white televisions and was renamed the Qingdao Television Factory. By the 1980s, the product line had further expanded to include colour TVs. It was during this period that the firm began cooperating with foreign firms such as Matsushita in order to learn advanced technology. In the 1990s, the firm broadened its product line into air-conditioners, computers, software, telephones, VCDs, DVDs, cameras, and optical and electrical products. The expanded company was restructured in 1994 to form the Hisense Group Co., which embraced a growing number of subsidiaries. In 1997, Hisense Electric Co. Ltd was listed on the Shanghai Stock Exchange. In 1999, Hisense won recognition from the State Administration for

Industry and Commerce of China (SAIC) as a producer of 'Well-known Branded Products in China'. In 2001, Hisense also became one of the five winners of China's National Quality Management Award from the National Association of Quality. Furthermore, Hisense is the only firm in the CE and home appliance industry to win the National Quality Award for four successive years. By 2004, Hisense had grown into a leading State-owned company with more than 7,000 employees. It has a presence in every major continent and sells its products to more than 100 countries in over 400 varieties (*www.hisense.com*).

The main business in Hisense is televisions, which account for 12 per cent of the domestic market share, ranking the firm No.1 in terms of the national market share in 2007 (*Chinese Household Electronics Report*, various years). The growth rate of its main business sales was 25 per cent in 2007, far above the average growth rate of 17.2 per cent in this product range (*Hisense Annual Report*, 2008). In the face of the fierce competition in the market, Hisense has benefited from market-oriented reform on the one hand, but on the otherhand, unrelenting pressures from competitors have reduced the profit margins of the company (Wu and Tang, 1999). To address the increased pressures, Hisense has revamped its business strategies and restructured the company.

Case 2: Nanjing Panda Group

Founded in 1936, the Nanjing Panda Group has experienced success and crisis in the recent past. As the earliest electronics enterprise in China, Panda used to be a large and leading SOE in the Chinese electronics industry. Its brand name was the first to be honoured by the State government as a 'China Famous Brand' in the Chinese electronics sector and was also the first brand name to enter the international market (*www.panda.cn*).

Two factors contributed to the success of Panda. First, Panda was a leading SOE in terms of the scale and importance of the sector. As a producer of defence-related electronics, the enterprise easily obtained the support of the government for both loans and other resources. Second, product shortages in China in the 1970s and 1980s created a golden opportunity for Panda to be a first mover in the CE sector from the end of the 1970s (Li, 2004).

However, the transition towards a market economy has brought huge institutional change in China. Government intervention has diminished and firms have great autonomy in deciding product type to price setting. Attracted by a surge in demand for CE products, a wide range of firms has entered this market. The company has struggled to adapt to the increasingly competitive market, which culminated in great losses for two consecutive years, 1997 and 1998 (Panda Electronics, 2002). The sales of its principal business decreased from RMB 2.54 billion (about US$309 million) in 1997 to RMB 1.43 billion (about US$174 million) in 1999 (Panda Electronics, 2002). In 1999, the stock of the company was classed as one with special treatment (ST Panda) by SSE, a euphemism for government intervention and support, and it faced being delisted from the stock market. Panda's market ranking dropped from 9.7 per cent in 1995 to 5 per cent in 2007 (*Chinese Household Electronics Report*, various years). The negative 10.44 per cent growth rate of Panda's main business sales in 2007 disabled it from competing with other manufacturers (*Panda Annual Report*, 2008). The fall in Panda's profitability was linked in part to excessive competition in the market and the company's ensuing inability to advance the development of original products offering high added value. Selling non-cutting-edge and out-of-date products made it uncompetitive with other domestic firms.

Case 3: Huadong Electronics Group Company

Huadong Electronics Group Company (hereafter referred to as Huadong) grew out of the National Second Electric Appliance Factory that was established in 1937. In July 1938, Huadong succeeded in producing the first incandescent lamp in China. Sixty years later, Huadong had become one of the biggest production bases of electronic vacuum devices and electronic lighting products with total assets of RMB 1.27 billion (about US$153 million) and 4,929 employees (including 1,283 engineering technicians and 142 senior technicians). The first Chinese incandescent lamp, fluorescent lamp, black-and-white picture tubes, colour picture tubes, monitor tubes, cathode ray tubes, counter tubes, photomultiplier tubes and infrared tubes were all produced by Huadong. Its current main products include more than 300 varieties in more than 20 categories including fluorescent lamps, compact fluorescent lamps, phosphor, filaments, electric glass, special electric lighting, electronic ballast and medical equipment vacuum display devices (*www.hdeg.com*).

Huadong has experienced a turnaround from crisis to recovery. As a SOE for more than 70 years, the firm was detached from the market for decades, but now has acknowledged the importance of the market (Li, 2001). This has enabled the firm to become more responsive to market conditions. On 20 May 1997, A-shares of Huadong went public on the Shenzhen Stock Exchange. Ten years after its IPO, the production capacity of the fluorescent lamp, the leading product, was more than 50 million pieces per year and covered 20 per cent of the market share by 2007, ranking as No.1 in the field of Chinese fluorescent lamps for its production capacity and sales quantity (*Xinhua Financial Network News*, 20 November 2003). The growth rate of its main business sales, 23.1 per cent in 2007, was also in the leading position in China compared with a 15 per cent average growth rate in similar product ranges (Huadong Annual Report, 2008).

Case 4: Haier Group Company

The history of the Haier Group Company dates back to 1984 when the current CEO, Zhang Ruimin, an official with the Qingdao municipal government at that time, was asked to take charge of the Qingdao General Refrigerator Factory (Palepu et al., 2005). When Zhang took over the management, the company was on the brink of bankruptcy, with no funds to pay the salaries of its employees or to invest in new product development. Over the past 25 years, the company has witnessed significant prosperity and is now a multinational enterprise.

In 1985, Haier imported technology from a German firm and began to manufacture technically sophisticated refrigerators. In 1989, the company changed its name to Qingdao Refrigerator Co. Ltd, and was restructured with funds raised from banks and government agencies. In 1991, the company once again changed its name, to Qingdao Haier Group Co., and in the same year it merged with Qingdao Air-Conditioner Plant and Qingdao Freezer General Plant. In 1992, the company set up Qingdao Freezing Equipment Co. and also merged with another previously State-owned enterprise, Qingdao Condenser Factory, which manufactured refrigerator condensers. In the same year, it became the first company in China to get ISO 9001 certification. The company's name was changed to the Haier Group in 1992 (*www.haier.com*). In 1993, Haier issued an IPO of RMB 50 million (about US$8.7 million) and was listed on the Shanghai Stock Exchange (SSE).

During the mid-1990s, Haier began to grow through mergers and acquisitions. In 1994, with revenues of RMB 2.54 billion (about US$317 million) and 10,000 employees, Haier was already known as one of China's best manufacturers of household appliances (Joseph, 2006). In the 10 years from its founding in 1984, Haier had grown from a tiny, failing neighbourhood workshop to become China's No. 1 refrigerator manufacturer, both in terms of quantity and with a reputation for the highest quality. Between 1995 and 1997, Haier acquired seven companies and started exporting its goods to foreign markets (Crawford and Feng, 2000). By 1997, Haier was the No. 1 consumer appliances brand in China and the market leader in all its product segments, which included refrigerators, washing machines, microwave ovens and freezers. Its revenues were reported at US$1.15 billion. In 1999, it opened its sales and marketing division in the USA. By 2002, Haier had a presence in 162 countries (to which it exported its products) in 42 categories with around 9,000 specifications (Khan, 2005).

Haier is recognized worldwide. In 2008, Haier ranked thirteenth on Forbes' Reputation Institute Global 200 list. Also in the same year, the company ranked first among Chinese enterprises on the *Financial Times* list of the most respected global companies (Duysters et al., 2008). Its main business of refrigeration products accounted for a 25.61 per cent share of the domestic market and it was the largest Chinese refrigeration appliance company by 2007 (*Chinese Household Electronics Report*, 2008). Compared with the industry's average growth rate of 12 per cent, the growth rate of Haier's main business sales, at 21 per cent (*Haier First Quarter Report*, 2009), shows Haier's transformation from a small, almost bankrupt enterprise to becoming one of the leading CE appliances makers in China.

Case 5: Chunlan Group Company

Chunlan started as a small air conditioner plant in 1985. Chunlan owned 42 subsidiaries worldwide and had total assets of 12 billion yuan by the end of 2007 (about US$1.4 billion). Located in Taizhou, Jiangsu province, it had a total of 8,475 employees. Over the past two decades, it has developed into a multi-functional international company, enlarged its industrial scale 600-fold, and increased its capital assets 700-fold and its economic benefits 500-fold (*www.chunlan.com*).

The company has introduced world-leading technologies, built an air conditioner R&D testing centre in China and has an annual productive capacity of 3 million units of various air conditioners (*Remin Daily*, 29 November 2000). National awards received include 'Chinese Top 10 High-tech Enterprises', 'Chinese Best Ten Joint Ventures' and 'Key High-tech Enterprise'. It has been one of the 'Chinese Top 500 Manufacturing Enterprises' for many consecutive years, and is identified as a knowledge-intensive and technology-intensive enterprise by the Science and Technology Department of Jiangsu province (Tang, 2002). In 2005, Chunlan realized an economic profit of 1.36 billion yuan (about US$164 million), accounting for a quarter of Taizhou's total. Its main business, air conditioners, occupied 12 per cent share of the domestic market in

2007 (*Chinese Household Electronics Report*, various years). This share ranks No. 5 in China (*Wenhui* newspaper, 28 August 2006). The growth rate of its main business sales of 30.27 per cent was much higher than the average rate of 13.84 per cent in 2007 (*Chunlan Annual Report*, 2008).

Case 6: Zhejiang Yankon Group Company

Zhejiang Yankon Group Company Limited was established in 1975. Like Tiantong (see below), Yankon was registered as a TVE (Chen, 1999). Yankon Group mainly manufactures energy-saving products including integrated electronic energy-saving lamps. Yankon is one of the earliest enterprises to have obtained the rights to import and export as an enterprise rather than via a government agency. It has the largest manufacturing base for producing and exporting energy-efficient lighting in China and is the largest manufacturer of this kind in Asia. Yankon cooperated with Philips and established the Zhejiang Yankon Lighting Co., Ltd, which has a dedicated R&D centre for energy saving products that recruits post-doctoral students to conduct further research into lighting products (*www.yankon.com*). In 2000, Yankon shares were listed on the SSE, the first company in the electric light sector to issue A-shares. In 2005 Yankon Lighting Electric Appliance Science and Technology Industrial Zone was started (*Shangyu Daily*, 27 June 2005).

As the most important Chinese lighting manufacturer in sales revenue and market share, the firm far out-competes General Electronics and Philips in the domestic market. Its products have obtained a series of international standard certifications: UL (Underwriters Laboratories Inc., US), FCC (Federal Communications Commission), VDE,[5] CE,[6] GS (Geprüfte Sicherheit, Germany), CSA (Canadian Standards Association) and ISO 9001 quality guarantee system. Yankon's energy saving lamps account for 15 per cent of the domestic market and the firm was the leader in this field in 2007 (*Chinese Household Electronics Report*, various years). A high relative growth rate of 26.10 per cent in 2007 in sales made it stand out above the domestic manufacturers relative to the 15 per cent average growth rate (Yankon Annual Report, 2008).

Case 7: Hangzhou Silan Microelectronics Company

Hangzhou Silan Microelectronics Company Limited (hereafter referred to as Silan) is located in the Hangzhou Hi-Tech Industry Developing Zone, Zhejiang province. Silan was founded by seven technological entrepreneurs in 1997 when the legality of the DPOEs became firmly established. Unlike Tiantong and Yankon, Silan was not originally registered as a TVE so as to disguise its identity (Xiao and Pei, 2003). In October 1997 Silan acquired a 40 per cent stake in the Taiwanese-invested Hangzhou Youwang Electronics Co. Ltd, in which the seven founders had once worked. The acquisition was via the founders of Silan transferring their shares in Youwang to Silan, after which Youngwang was re-registered as a joint venture company of Silan and the Taiwan-based Unisonic Technologies Co. Ltd. Silan specializes in designing, developing and manufacturing integrated circuit (IC) products for applications in communication, network and digital household electric equipment (Listed Company Association, 21 March 2003). In September 2000, Hangzhou Silan Electronics Co. Ltd was renamed as Hangzhou Silan Microelectronics Joint-stock Co. Ltd, which in 2005 had a total capital of RMB75 million (about US$9 million) (*South China Morning Post*, 31 August 2005).

After several years of rapid development, Silan is ranked in the forefront of its counterparts in China in terms of technical level, business scale and profit-making ability. It has grown into one of the major designers and providers of civil IC products in China. According to CCID (China's Centre for Information Industry Development) data in 2005, it was the second largest domestic IC design house with 12 per cent share of the market (Listed Company Association, 2 July 2007). In 2007, its domestic market share reached 20 percent, and it became the largest IC manufacturer (*Chinese Household Electronics Report*, various years). With the rapid increase of its market share, the growth rate of Silan's main business sales of 39.16 per cent was also ahead of the average level of 19.2 per cent in the domestic market in 2007 (Silan Annual Report, 2008).

Case 8: Zhejiang Tiantong Electronics Company

Established in 1984, Tiantong Electronics Company Limited (hereafter referred to as Tiantong) has developed as a key manufacturer in soft magnetic products in China, which are widely used in communication equipment, computers and peripherals, electronic and automatic control instrumentations, TVs, displays and auto electronics (*www.tiantong. com*). Although privately owned, Tiantong was registered as a TVE until 1999, which was a means overcome political discrimination against the private sector (Cheng and Ma, 2006).

In 2000, Tiantong was listed on the SSE. It is the first Chinese listed company to have been controlled by a natural person,[7] with more than

3,000 employees and up to RMB 439.15 million (about US$53.6 million) registered capital (Guan et al., 2001). Experts at home and abroad had pointed out that Tiantong's listing was an important sign that listed firms in China had become much closer to the international standard (*www .sina.com*, 15 July 2003).

After more than 20 years of development since its establishment, Tiantong has become a recognized leading provider of electronic materials and components. The firm is ranked among the top 100 enterprises in China, a key enterprise in Zhejiang province and an ISO 9002-certified company (Lv, 2004). The products of the firm accounted for 35 per cent of the domestic market in 2007 and it is the largest firm in terms of its market share in this field in China (*Chinese Household Electronics Report*, various years). The competitiveness also made Tiantong take the lead in the growth rate of its main business sales of 26.30 per cent compared with the average level of 15 per cent (*Tiantong Annual Report*, 2008).

Conclusion

This chapter has brought together a review of the overall institutional setting in China and thus the chapter fulfils the objective of reviewing its evolution and the current institutional setting. In doing so it highlights the complexity of the context in which this study falls and provides a broad understanding of the background, evolution and current standing of reforms in China. The profiles of the case study firms provided a concise historical overview of the development of the firms.

Overall, during the past three decades China has undertaken a dramatic transition from a planned to a market-oriented economy, in which the influence of market forces on the allocation of resources in various economic spheres has been significantly enhanced. This is a new development, which is different from both Western market economies and formal planned economies (Liu and Garino, 2001a). One of the most important dimensions of economic transition is that the reformed institutional arrangements in China are conducive to sustained economic growth.

Notes

1. Under this system, different enterprises belong to either central industrial ministries or local governments. There is a lack of horizontal flow of factors of production and efficient allocation of resources, which undermines the

efficient allocation of resources and results in serious imbalances and shortages (Guo, 2007).

2. The first step of corporatization was to separate non-productive assets such as schools and hospitals from productive ones. Productive assets account for 50 to 75 per cent of total assets of the to-be-listed stock companies, while non-productive assets were left with the SOEs. All retired workers also remained on the SOEs' payrolls. An accounting firm was then hired to audit the financial statements of the SOEs for the last three years and separate productive assets. In the meantime, managers of the SOEs contacted other enterprises and institutions to see if they were willing to be legal co-founders of the stock company. The SOEs also talked intensively with the local government and party for candidates for managers and board members (Xu and Wang, 1997).

3. The China Securities Regulatory Commission was established in 1992 with the approval of the State Council of the People's Republic of China as the executive agency of the Securities Policy Committee of the State Council in the supervision and regulation of the national securities and futures markets in China (*http://www.csrc.gov.cn*).

4. The IPO is designed for domestic private investors. The basic idea is that an old SOE carves out its better performing assets to create a subsidiary through a process of 'packaging', and invite private investors to contribute additional capital by issuing new shares. The most important feature is that domestic private investors are mostly individuals. As a result, corporate control stays with the State while private equity capital enters the firm. The non-tradability of State-owned and legal person-owned shares provides further assurance that State-owned shares are not sold to private investors in the market (Mako and Zhang, 2003). On the day of the IPO, at least 25 per cent of total shares are sold to the public (Chinese Company Law, 2005). After the IPO, the original SOE either disappears or becomes the majority holder of the listed firm (Xu and Wang, 1997).

5. VDE is the German Association for Electrical, Electronic and Information Technologies, a professional national standards and test agency.

6. CE is a mandatory European marking for certain product groups to indicate conformity with the essential health and safety requirements set out in European Directives.

7. In China private enterprises are often invested and established by another private company or by natural persons, or controlled by natural persons using employed labour. Included in the category of private enterprises are private limited liability corporations, private shareholding corporations and private partnership enterprises, and privately funded enterprises registered in accordance with the Corporation Law, Partnership Enterprises Law and other regulations on private enterprises (*www.adbi.org*). The minimum registered amount for a natural person is no less than RMB 100,000 (about US$ 14 628), which must be paid up in just one instalment according to the law (Chinese Company Law, 2005).

4

Business strategies in the focal companies

Movement from a centrally planned to a market driven economy changes fundamental managerial assumption, criteria and decision-making, and represents a genuine transformation of the business.

(Justin Tan and Robert Litschert)

Abstract: A firm's specific strategy selection is based on the careful evaluation of its unique resource portfolios. Two distinct ways of developing the resource and capability base of a firm are exploitation and exploration strategies. Informed by the literature on the institutional perspective and using the exploitation–exploration framework from Chapter 2, this chapter examines the coevolution of strategies with changes in the industry and the overall Chinese business environment.

In a transition economy like that of China, how do exploitation and exploration strategies apply to firms? How have Chinese firms achieved a dynamic strategic fit between their changing environment and their resources and capabilities? We address these questions by drawing on insights from a multi-case method. Analysis of within and cross-case data from eight Chinese firms in the consumer electronics sector revealed two major findings: First, institutional transformation exerts a strong influence on the adoption of business strategies. Environmental changes require changes in firms. The specific content of the strategy coevolved with changes in the industry and the overall Chinese business environment. Second, the competitive position is a matter of the degree of fit with the environmental dynamics that firms have been able to realize. The firms whose activities have achieved an appropriate fit with the environment are successful, while those unable to adjust to the changing market demands have lagged behind. Our findings contribute to research into business strategies in transition countries.

Key words: exploitation, exploration, strategy, China, case study.

Introduction

This chapter elucidates the business strategies of Chinese companies in the CE sector. These business strategies are discussed on the basis of resource exploitation strategies and resource exploration strategies for future development (March, 1991). Firms base their strategies on how to best use existing core resources and capabilities in order to survive and thrive, which we call exploitation strategy. In addition, during the transition, firms call for strategies to develop new resources and capabilities in order to compete in a highly competitive sector, which we call exploration strategy. An important feature of the cases in adopting business strategies is that they exhibit a different choice between exploitation and exploration at different stages of development. This choice is designed to cope with institutional changes, which include market turbulence and social transformation (White and Xie, 2006).

The development of China's firms has evolved through two broad stages with making a turning point in 1993. In that year, the CCP adopted the Decisions on Some Problems in the Establishment of the Socialist Market Economic System, which shifted the focus of enterprise reform from the devolution of managerial authority from the state to the enterprise, to a focus on a market-oriented enterprise system (China Internet Information Centre, 2004, accessed at *www.fdi.gov.cn*). Changes in the environment have had a significant impact on the strategic choice of the firms. The cases describe a development process by which the firms deepen their existing set of resources. The cases also show how each stage in the firms' development of resources and capabilities required different strategies for maintaining a fit.

This chapter also compares different strategic choices of firms responding to the changing opportunities and constraints inherent in their environment. Thus the cases provide insight into the interactions between environment changes and managerial decisions that influence firm performance, which have both theoretical and empirical implications. Theoretically, it explores the exploitation vs exploration strategies in

China's transition context. Empirically, the experiences of these companies will shed light on the reform of the firms in China.

This chapter divides the strategy of each category of firms into two stages with 1993 as the watershed. Sections 2, 3 and 4 provide an analysis of business strategies of the SOEs, COEs and DPOEs respectively based on their resource exploitation and exploration. Section 5 explains the results of the strategic choices. Section 6 concludes the chapter.

The development of exploitation and exploration strategies in SOEs

The development of exploitation and exploration strategies at Stage 1 (1978–1992)

Institutional factors influencing the adoption of strategies

Since the initiation of economic reforms in the late 1970s, China has achieved impressive economic growth coupled with significant structural transformation. China started economic reform in 1978–1984. The government put forward the idea of 'delegating power and sharing profits' with enterprises for the purpose of carrying out reforms. By 'delegating power and sharing profits', the government intended to bring forth the initiative of SOEs in production and operation (China.org.cn, 7 November 2003). From 1985 to 1992, the efforts of the reform concentrated on giving the SOEs more autonomy by allowing them increased authority over the allocation of their profits, and limited production autonomy. Central government decided to turn SOEs into independent production and management entities, and stipulated that enterprises take responsibility for their own profits and losses (Naughton, 1995). A comparison of the government quotas and the aims of government control with actual performance in industrial growth clearly indicated that most of the government quotas had lost their binding power. The development of the firms started to be guided by the market mechanism.

During this period, SOEs still played a dominant role in the CE sector in which more than two-thirds of the top enterprises in terms of total output were SOEs (Jiang, 2001c). Since the mid-1980s, although the SOEs still enjoyed many privileges including receiving government grants or loans for their expansion, guaranteed markets for their products, and ready access to raw materials, the government's support for the SOEs in the CE sector declined (Huchet and Richet, 2002).

From the mid-1980s the government allowed international firms easier access to the Chinese market. However, Chinese consumers preferred local brands, largely for reasons of price (Interview 3). Domestic firms in the CE sector began to gain market share because of the price advantage and improving economies of scale (OECD, 2002). At the end of the 1980s, the domestic market for CE products shifted from one of shortage to over-supply. The State-owned distribution organizations were highly rigid and low in efficiency for manufacturers (Interview 3). Consequently, the firms could not respond promptly and flexibly to changing customer needs. The increasing misfit led some firms in the CE sector to adopt corresponding business strategies to rectify the problems in the economic system.

The adoption of resource exploitation and exploration strategies

Resource exploration

During this stage, Hisense and Huadong began to build their market knowledge by interacting with customers and creating extensive distribution networks. Access to a distribution network was a scarce and competitively valuable resource in the first stage of firms' development (Interview 5).

Under the central planning system in economies such as China, the sale of all products was highly centralized and the companies did not have their own distribution outlets. State-owned distribution organizations were found in all industrial sectors, which were legacies of the central planning system responsible for fulfilling the allocation directives of the State Planning Commission and its relevant industrial bureau for manufacturing inputs, intermediary products and final goods (White and Xie, 2006). Changes to the distribution system only began in the mid-1980s. Manufacturers started to bypass State distribution organizations and sell directly to retailers or even directly to consumers. In the belief that the technology inherent in the products of all the domestic firms was similar, distribution became the key focus of competition, and was the companies' most important point of differentiation (Interview 53). The distribution routes owned by companies achieved two positive outcomes. First, unlike State-run distribution activities, the distribution channels owned by the companies themselves were consumer-focused by necessity. Management realized that the companies could only survive by matching with consumer demand (Interviews 3, 5, 17, 19). Self-owned distribution channels encouraged companies to generate improvements in quality, pay greater attention to consumer needs, undertake better warranties and repair services, make greater efforts at sales promotion, and develop new product varieties to meet market needs. Second, a strong commitment to understanding and working with this distribution system helped prevent problems that arose in different regions, and symbolized a commitment to the Chinese market. As one senior manager from Huadong claimed:

> Contrary to the developed market where firms must rely on research and development and strong products to be competitive, in China the technology inherent in the products was similar. We began to build up our understanding of Chinese consumers and their purchasing habits through creating our own distribution network. Thus the advantage of the distribution channels explained to some extent the good performance of some firms. (Interview 19)

Hisense and Huadong were two of the earliest companies in the sector to build a nationwide system of provincial sales offices. Senior management from the two firms believed that during the early stage of development creating a distribution network to acquire market knowledge was more practical than developing advanced technology independently, as the latter required heavy investment and faced high risk (Interviews 3, 5, 16, 19). The sales network of Hisense and Huadong reduced their distribution costs and enabled them to compete better on price. Toward the end of this stage, Hisense and Huadong became the leaders in their respective areas. Before 1991, Hisense was a relatively unknown firm. The second year following the establishment of its own distribution channels, the output of its main business, colour TV sets, exceeded 1.4 million units, ranking the firm fourth among the TV producers in China (Interview 3). Huadong also developed into a nationally well-known enterprise, producing more than 300 kinds of products by 1992 (Interview 20). This strategy boosted Hisense and Huadong's image in the market.

Resource exploitation

Hisense and Huadong broadened their R&D activities in order to support their marketing activities. The R&D activities in this stage mainly focused on cooperating with MNCs. As early as 1984, Hisense had entered into a collaboration with Matsushita to acquire colour TV technology (Interview 1). Subsequently, other collaboration agreements were signed between Hisense and foreign companies. Collaboration took various forms, from buying patents and licensing to establishing joint ventures. Although buying technology from foreign firms was a common practice, Hisense bought only advanced technologies from Western countries. Senior managers of Hisense considered that this would be more effective when competing with local firms and in catching up with MNCs. When facing the choice of buying technology from a Hong Kong firm for US$1.5 million or from Matsushita for US$3 million in 1992, Hisense chose to buy from Matsushita, which had more advanced technologies (Interviews 3, 5). These collaborations with MNCs enabled Hisense to acquire the technology within a short time and reduced the stigma of lagging technology attached to local brands by Chinese consumers.

Huadong lacked experience in advanced information technology. The firm realized that it would be time-consuming to develop technology independently and the cost would be colossal. The senior management of Huadong decided to cooperate with foreign and domestic companies that had advanced technology (Interviews 16, 20). Huadong formed strategic alliances with the following domestic companies:

- Nanjing Sanbao Industrial Company Ltd was a privately owned enterprise whose digital electronic products had great potential. However, this company was constrained by a lack of capital. In 1992 Huadong made a cash investment of RMB10 million (about US$1.2 million) and Sanbao supplied its intangible assets and equipment to jointly establish the Nanjing Information System Co. Ltd.
- Huadong also established 10 joint ventures with potential for high efficiency and growth, such as Huafei Color Display Systems Co. Ltd, Feidong Lighting Co. Ltd, Nanjing Electronic Shadow Mask Co. Ltd and Nanjing Saes Huadong Getters Co. Ltd. These allowed Huadong to improve their advanced technique between 1986 and 1990.

Moreover, Huadong also cooperated with foreign firms such as Philips and General Electronics (GE) in the form of original equipment manufacturing (OEM) arrangements. The senior management of Huadong considered that OEM contracts would serve as training programmes for the company: through the OEM process, Huadong accumulated assembly and process engineering knowledge from suppliers of equipment and technical assistance from contractors, and improved its capabilities through learning-by-doing (Interview 16). As a business partner of foreign companies such as Philips of Holland, Thorn of the UK, and GE, Cooper and Thomas of the US, Huadong believed that OEM would lead the firm to concentrate on manufacturing rather than on developing technology. Huadong's use of OEM differed from others (Interview 19). Chinese firms commonly provided cheap labour, while the business partners supplied other resources including technology. However, Huadong combined external and self-owned technology in OEM, which enabled the firm to make larger profits (Interviews 19, 20).

Unlike Hisense and Huadong, Panda was less eager to cooperate with firms from advanced countries for technology and paid less attention to distribution channels. Panda was instead more interested in developing cutting-edge technologies internally that would enable it to enter the high-end market on the basis of its own innovative capability (Interview 9). For example, in the early 1990s, large-screen (29-inch) TV sets in China were high-technology products. Panda tried to develop chips for such large-screen TV sets independently (Interview 9). However, this strategy was unsuccessful. Owing to the weaknesses of the technology inherent in its products, Panda's TV sets of this kind were squeezed out of the market, resulting in a loss of market share to its competitors who used imported technology. The senior management of Panda realized too late the weakness of this strategy and their urgent attempt to import technology failed (Interviews 8, 9).

The development of exploitation and exploration strategies at Stage 2 (1993–2007)

Institutional factors influencing the adoption of strategies

The year 1993 was a watershed for the conversion of China's SOEs. Since 1993, both central and local governments had become less directly involved in commercial activities and further deregulation of production and circulation of products was achieved. Accordingly, the manufacturers had more freedom and greater willingness to innovate according to market liberalization, competitive pressure, and consumer demand (Jefferson and Su, 2006; Song and Yao, 2003). Increased market competition, globalization and technological innovation had reshaped the competitive landscape of the surviving SOEs (Tan, 2005). The overwhelming majority of interviewed respondents from the SOEs agreed that the Chinese government had begun to relinquish control over SOEs in competitive industries and to increase market discipline on them during the 1990s.

On the supply side, companies were permitted to procure supplies directly from suppliers when their needs exceeded their allocation. For example, the interviewees from Hisense and Huadong said the company had considerable freedom to plan and arrange their activities. The management had gained relative independence from the local government (Interviews 1, 2, 3, 18, 19, 20). The Qingdao and Nanjing municipal governments allowed the firms to sign sales contracts directly with commercial units at their own discretion, and did not impose burdens on Hisense and Huadong that could harm their efficiency and profitability. Hisense and Huadong could also make their own investment and financial plans (Interviewees 1, 3, 5, 18, 20). In this way motives for the firms to maximize profit had been mobilized.

The senior managers of the three SOEs generally acknowledged that they had more authority over R&D activities, product innovation, investment planning and marketing during this period (Interviews 2, 6, 8, 16). Increased managerial autonomy allowed firms the freedom to adopt innovative measures of their own in order to improve the performance of the firms. One manager of Huadong commented:

> We have gained more autonomy and incentives. Since investment in advanced technology and quality assurances is the key to competition, we have to invest heavily to upgrade production technology. (Interview 16)

Adopting resource exploitation and exploration strategies

Resource exploitation

During this stage, senior managers from both Hisense and Huadong believed that independent distribution channels were still crucial resources for them. They continued to develop their distribution network, which had given them increasingly greater geographical coverage compared to other domestic producers. By the end of 2006, Huadong had established subsidiaries throughout China and could boast 5,200 direct selling outlets and intermediate agents (Interview 20), while Hisense had 200 branches and more than 10,000 sales and service outlets throughout China covering of all provinces and major cities:

> Compared with other companies, our [Hisense] colossal market
> network is an advantage we may be proud of. Although there is
> competition among distributors, Hisense is able to maintain a
> positive relationship with its distributors, many of which have
> grown with Hisense over the years. (Hisense's CEO, Zhou Houjian
> cited in *Asia Port Daily News*, 4 January 2000)

Hisense and Huadong's further expansion of their distribution network
and sales-and-service activities supported marketing activities and
product design decisions. The firms incorporated feedback and the
experience of users obtained from its distribution channels and marketing
departments into product design and innovation efforts in its business-
level R&D centres (Interviews 2, 6, 16, 18, 19).

Hisense and Huadong pursued technological innovation through both
importation and cooperation. International partners and 'buying in'
became the main sources of learning. There were two reasons for the
choice. First, many of China's firms still had limited capability for an
in-house technology development during the second stage from 1993, so
they relied on international outsourcing of some technology (Interviews
53, 54). Second, the senior management of Hisense and Huadong believed
that some in-house R&D was a long-term oriented activity the future
outcome of which was uncertain (Interviews 1, 17, 20). In the 2000s, they
still could not afford the huge expense of the technology development
costs. The incessant demand for fresh products, the attempt to telescope
product cycles, and the extraordinary costs of pushing the technological
frontiers are enormous challenges. Because of the breadth of technologies
and capabilities relevant to CE, the managers of the two firms recognized
that they must supplement internal R&D activities, especially those
targeting the future (Interviews 53, 54).

Resource exploration

The management of firms identified technology and innovation as the
focus of their new strategic development. Acquiring new resources,
especially knowledge, and creating varieties in innovation are important
processes inherent in business strategies. These were essential to grow
the firms since the CE sector was characterized by rapid changes in
product features, functions and performance (Interview 53). From the
mid-1990s, Hisense and Huadong had embarked on establishing an
internal R&D capability in addition to the import of technology. Since
over-production of the consumer electronics products had forced the

market price down over the past few years to the detriment of producers, Hisense and Huadong felt that they could demand a price premium that was good for the profits (Interviews 1, 19). Rather than being content to play in the conventional business segment, they were seeking the higher margins of new technology and making a leap forward in consumer electronics through technology. Their first attempt was to lay a solid foundation for the manufacturing of high-tech products. Hisense and Huadong invested in manufacturing capabilities and establishing large-scale manufacturing plants. Hisense's Technical Park had more than 1,500 staff, 11 professional institutes, five associated research centres and a postdoctoral research and development working station by the end of 2007. In the early 2000s the company also located research facilities in Japan and the USA, where the world's cutting-edge technology was being developed (Interview 6). Huadong had also been able to build one of the most modern technical centres in China (*http://www.hdeg.com*).

The two firms increased spending on R&D, which was mainly directed towards testing and quality enhancement rather than towards the development of radically new technology. At Hisense for example, its R&D investment in innovation heavily emphasized utility model and design patents instead of original innovation (see Figure 4.1). The development of innovation capabilities had helped make Hisense one of the leading local firms in the TV set industry.

Panda was confronted with the challenges to its survival in the market due to the strategic errors. It was obvious from 1996 that Panda had lost its advantages in TV manufacturing and was quickly losing its position

Figure 4.1 Percentages of the three kinds of patents in Hisense, 2005

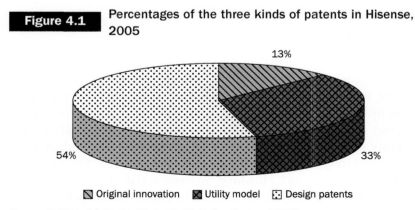

☒ Original innovation   ☒ Utility model   ☒ Design patents

Source: Chinese Intellectual Property Statistics Bureau (2005).

| Table 4.1 | Leading brands in China's CTV market (1996, 2007) |

1996		2007	
Leading brands	Market share (%)	Leading brands	Market share (%)
Changhong	20.5	Hisense	16.23
Konka	12.2	TCL	14.29
Beijing	7.1	Skyworth	14.21
TCL	6.2	Konka	12.23
Panda	4.6	Changhong	6.05

Sources: Joseph (2006); Interviews (2, 6, 7).

in the face of competition from more dynamic firms (see Table 4.1). In 1998, Panda manufactured 1.3 million TV sets, while its unsold stock exceeded 400,000 sets (Interview 8). Hisense by 2007 had taken over Panda's position as the leading TV maker in terms of market share (see Table 4.1). Eventually, due to its lack of motivation to make changes to adapt to the changing market, Panda lost over one-third of its market share in China (Interviews 8, 14).

The government had made great efforts to help bail out Panda to ensure its survival. In order to improve its performance, the deputy mayor of Nanjing Municipality (the host municipality of Panda Group) was appointed chairman of the Panda Group in 2003 (Li, 2004). Again, the comments of one interviewee from the Panda Group were illustrative:

> You cannot imagine the intricacy of relationship web of Panda in the government. Panda is the only company in China that has been visited by each President of China since the 1950s: Mao Zedong, Deng Xiaoping, Jiang Zeming and Hu Jintao. It will be unbelievable if Panda, one of the earliest and ever most famous SOEs is closed down. (Interview 10)

These visits and 'soft promotions' by State leaders were worth millions of dollars. In 1999, when Panda was heavily in debt due to a price war in the CE sector, the municipal-level and provincial-level governments provided RMB 500 million (about US$60.46 million) to assist the company with strategic restructuring (Yu, 2005). In April 2001, the Chinese President Jiang Zeming visited Cuba. During the visit, the two countries signed a credit agreement that included the export of Panda products to Cuba (*www.xinhuanet.com*, accessed on 28 February 2003).

The extent of the relationship networks that Panda managed to build up over time was one of its most striking features. The dependence of Panda on the government made it unresponsive to market changes. Originally Panda was the premier competitive domestic colour TV manufacturer in China; by the early 2000s its survival was no longer possible without government support. Yet with the company being one of the largest in the province and one of the earliest national brands in China, the government was reluctant to close it, despite continuing losses. The State pledged that every attempt would be made to assist Panda to overcome its financial plight (Yu, 2005).

The development of exploitation and exploration strategies in COEs

The development of exploitation and exploration strategies at Stage 1 (1978–1992)

Institutional factors influencing the adoption of strategies

During this period, COEs did not enjoy as many privileges as SOEs in terms of receiving subsidies, bank loans and scarce resources from the State, but they were subject to State controls and regulations (Interview 25). In the early 1980s, there were indicators that the urban collectives had been discriminated against in many aspects by the government. The COEs appeared as a second-class, less attractive alternative for employing millions of people whom the SOEs could not absorb (Li, 2004). The lack of privileges from the government at their early period of development forced them to learn earlier how to compete in the market. Chinese banks preferred to lend to established firms rather than new entrants. Older firms would receive more favourable treatment from the banking system than younger ones (Haggard and Huang, 2008). Thus the growth of the COEs during this early period had resulted largely from the use of local resources without much financial assistance from the central authorities. According to a board director of Haier:

> The SOEs were able to get access to many privileges in the early-1980s. The government assisted them in terms of land and

investment. Haier was only a small handicraft factory then, we did not benefit from the State's largesse. We had to obtain a loan from the bank in order to buy the production line. (Interview 22)

As with Haier, Chunlan was in a similar situation. Chunlan emerged in the mid-1980s as an amalgamation of three small township collectives. At the time there were fewer than 1,000 employees, and fixed assets of RMB 3.5 million (about US$636,000). One manager stated:

It was very difficult for us to borrow even RMB 50,000 [about US$10,000] from the bank at that time. Chunlan was unable to get access to the privileges enjoyed by the SOEs. We were forced to plunge into the sea of the market. However, when we look back it was a good thing since it cultivated Chunlan's ability to survive independently. (Interview 28)

The adoption of exploitation and exploration strategies

Resource exploration

More effective channel structures in the CE sector began to emerge at this stage, which allowed firms to learn to set up intricate connection points between logistical and distribution nodes (Interview 53). The management of Haier and Chunlan believed that a market share would determine their competitive strength in the sector. Market knowledge and a high degree of operational flexibility could be achieved through business networks (Interviews 25, 28). By the early 1990s, Haier's distribution centres in the main cities of China were operating as independent sales companies that had to be responsive to the needs of consumers in order to remain profitable. The distribution strategy for its Chinese business was based on a 'Channel Building' team. Instead of establishing agent dealers in separate cities, it set up general agencies in provinces, which recruited regional managers. Each regional agency worked with several lower-level distributors, who in turn had sub-distributors that served exclusive regional territories. It enabled Haier to become the No. 1 producer of consumer electronics products in China (Interview 25).

Chunlan also developed strong domestic distribution channels. Chunlan's products first sold mainly in large and medium-sized cities in China and only later expanded distribution outlets to small townships (Interview 30). In second- and third-tier cities, Chunlan set up networks of licensed dealers that accounted for 30 per cent of the sales. Independent retail shops and government purchases accounted for another 15 per cent each, with telephone sales making up the rest (Xiao and Pei, 2003).

Resource exploitation

At this stage, Haier and Chunlan started to focus on the acquisition of core technology. The interviewees believed that Chinese firms have not developed their own core technologies and feared that MNCs from developed countries with advanced technology would enter the Chinese market, leaving no space and time for organic development by local firms (Interviews 22, 25, 27, 28, 30). Like their SOE counterparts, Haier and Chunlan would rather use a 'buy in' strategy than independently develop their own advanced technology. They had acquired the technology either directly, via a licence for example, or they allied with another firm in a partnership or joint venture (Interviews 22, 27, 28).

Haier and Chunlan were able to gain access to advanced Western technology through these alliances, which enabled them to produce appliances of higher quality. Technology alliances proved to be instrumental in Haier's strategy. Haier had alliances with Liebherr, Philips, Motorola, Merloni and Mitsubishi (Joseph, 2006). Chunlan's business partners included General Electric, Ford, Boeing, Motorola and the Mitsubishi Corporation (Interview 28). These alliances helped Haier and Chunlan in overcoming a major handicap that most Chinese COEs suffered from – the lack of adequate R&D. For example, in 1984 Haier introduced technology and equipment from Liebherr, a German company, to produce several popular refrigerator brands in China. Liebherr had 70 years of experience in producing high-quality refrigerators. Meanwhile, Haier expanded cooperation with Liebherr by manufacturing refrigerators based on its standards, which were sold to Liebherr for the German market. According to an interviewee, in this regard Haier's strategy was 'to introduce advanced technology and equipment, and assimilate it as rapidly as possible by way of imitation, and at the same time take the initiative to cultivate and train its own technological personnel' (Interview 50). Haier followed up the licensing of Liebherr's four-star technology with an active learning strategy, which included establishing a R&D department and sending more than 40 engineers and managers to Liebherr for training. Liebherr proved to be a successful training institute for Haier's R&D talents. In 1985, a year after it licensed Liebherr's technology, Haier was able to introduce its first four-star refrigerator to the Chinese market. This product established Haier as the leading refrigerator producer in China (Interview 50).

The development of exploitation and exploration strategies at Stage 2 (1993–2007)

Institutional factors influencing the adoption of strategies

With the 'modern enterprise system' experiment introduced in 1993, the number of small and medium-sized COEs declined, while the large and successful COEs received support from the government and gained easier access to external markets (Yin, 1998). Large-scale collectives such as Haier and Chunlan play a significant role in providing employment, supply products and contributing tax revenues to the local government. Assistance from the local government therefore helped to create a more favourable task environment for large collectives. Owing to their success, the firms have become important enough for the parent municipalities to be able to exercise little leverage on them (Yin, 1998).

Haier and Chunlan were operated with a high level of managerial independence during this period. Senior managers of the companies agreed they had great independence in making decisions (Interviews 22, 24, 26, 27, 28, 30). The decision-making style of the CEOs of both Haier and Chunlan was highly entrepreneurial. When it came to business strategies, the prestige and reputation of these two firms usually carried the day, so long as municipal officials believed their strategies to be sound. As one manager revealed:

> We are no longer dependent economic units and subordinate factories under immediate control of local cadres who assumed social responsibility for guaranteeing job opportunities to local people. As long as we could persuade the government our strategies were reasonable, the government would support our decisions. (Interview 26)

China had developed a series of indexes to measure the political performance of local government officials to determine their political future (Gao and Tian, 2006). Among these indexes, the most important was the local economy. Local government officials had more opportunities to be promoted if they had achieved a good economic performance (Interview 52). Haier and Chunlan are valued enterprises for local government. They were nationally famous, large-scale and well managed, and contributed greatly to the local government coffer. In order to attract or foster local large enterprises, or prevent this kind of enterprise from relocating, local government policies were favourably geared towards their development (Interviews 23, 29, 52, 54).

The adoption of exploitation and exploration strategies

Resource exploitation

Complementing their more differentiated product line, Haier and Chunlan continued to elaborate their distribution system in order to address more finely the geographical variations in customer purchasing power, lifestyle attitudes and consumption patterns. By 2004, Haier had a service workforce of 5,500 independent contractors. Haier's products also successively entered Europe, North America and other developed-country markets. Haier established a global network of design, manufacture and distribution services. Based on these networks, Haier met localized market demand at home and abroad with innovative models (Joseph, 2006).

Since 1997, Chunlan has established an overseas sales network and developed over 1,200 overseas agents. To acquire an overseas distribution network, Chunlan established strategic alliances with many international rivals so as to promote mutual benefits with their respective advantages and resources. For example, Chunlan found it very difficult to market its own-brand products in Japan. In 2002, Chunlan entered the Japanese market through an agreement with Sanyo Electric Company. The agreement resulted in the setting up of a joint venture (JV), Sanyo–Chunlan Co., which sold Chunlan's products. In return, Sanyo's products were sold in China through Chunlan's sales and distribution network (Interview 27, 28).

During the mid-1990s, Haier and Chunlan began to acquire weaker competitors in the market. This acquisition strategy sought to gain good manufacturing facilities and enlist trained employees with less capital expenditure (Interviews 22, 30). Haier and Chunlan would identify viable enterprises in specific product areas. These enterprises were usually on the brink of bankruptcy, but had good manufacturing facilities and well-trained employees (Interviews 26, 28, 30). With the further marketization in China, Haier and Chunlan were able to acquire SOEs as well as COEs. Although the acquisition of SOEs involved far more complex negotiations than that of COEs, SOEs held the promise of even better production facilities. The trick was to turn the companies around in cooperation with the relevant political authorities (Khan, 2005). One of Haier's most famous acquisitions was the Red Star Electric Appliance

Company in 1995, a major Qingdao-based manufacturer of washing machines, with over 3,000 employees but about US$300 million in debt. Red Star was in theory controlled by the municipal government and its workers, but Haier managed to gain almost complete freedom to convert the washing machine plants to its own methods. Haier had exploited the ambiguity of ownership laws in China to act as acquisition entrepreneur. The CEO of Haier Zhang Ruimin called this process 'hunting the stunned fish, not the dead ones' (Joseph, 2006). As a result of their acquisition strategy, Haier and Chunlan's growth in the following years was rapid.

Resource exploration

A strong capability to absorb acquired technologies and develop new products was critical to meet market changes. Haier and Chunlan had tried to develop their own capabilities to overcome barriers in implementing this strategy. The two case firms had been implementing a technology building strategy involving external sourcing of technology in combination with building internal capability through investment in R&D and design. In addition to the cooperative research programmes with leading foreign companies, they attempted to develop their own advantages instead of being solely reliant on their business partners. As an interviewee from Haier explained:

> The technology suppliers from Western countries are not always willing to disseminate core technology to us. The firm can only acquire some medium or low-level technology in this way. (Interview 22)

Before 1996, the proportion of Haier's R&D expenditures as a percentage of total sales was about 3 per cent; it subsequently reached 4 per cent in 1997, and climbed to about 5 per cent in the next three years before reaching 6.6 per cent in 2000. Although R&D spending dipped slightly as a percentage of sales (it continued to increase in absolute terms) it bounced back to over 6 per cent during 2005 and 2006. In 2006, 70 per cent of Haier's total R&D expenses were on investment in overseas R&D (Interview 24). Haier established 19 production factories and two 'production parks' outside China based on its own accumulated R&D capability (Kiran and Chaudhuri, 2004). Although subsidiaries were established after 1996 in nearby developing countries such as Indonesia, the Philippines and Malaysia, internationalization had been directed primarily toward developed countries like the US, Canada,

Japan, France and the Netherlands. The foreign subsidiaries enabled the company to develop home appliances that met the needs and wants of local consumers. Haier's strategy was to build an international brand name in the toughest developed markets, from which it could gradually expand to other markets (Interview 25). Haier's R&D spending appeared to match or even exceed industry standards. Table 4.2 indicates that Haier's spending on R&D as a proportion of sales revenue had been approaching that of Sony and was higher than GE's.

Chunlan adopted the strategy to consolidate short and long-term targets, emphasizing both technology research and application, and innovation. After years of effort, the company established a technology innovation system focused on short, medium and long-term prospects. It established a Central Research Institute in 1997 (Interview 29).

> In order not to be hindered by out-dated technologies, we have endeavoured to develop our own core technologies to shorten the gap with the international standard, and build an innovative approach with our own characteristics. (Interview 29)

Chunlan's research institute performs fundamental and prospective research and is focused on the latest technology. The centre consisted of three main levels. The first level was the corporate group, which was responsible for the development of core technologies and basic research. The second level was created in every department (business unit). The third one was connected to every plant (cost centre). The aim was to catch up with the world's leading technology within five to 10 years and integrate the development of new technology with the creation of new products (Interview 29). The establishment of Chunlan's tech-innovation system helped the company to succeed in transforming from 'import and

Table 4.2	**R&D spending as a percentage of sales, 1998–2006: A comparison between Haier, General Electric (GE) and Sony**

Company	1998	1999	2000	2001	2002	2003	2004	2005	2006
Haier	4.6	4.8	4.8	5.6	4.8	4.3	4.4	6.2	6.2
GE	1.5	1.5	1.5	1.6	1.7	1.6	1.6	1.8	1.8
Sony	5.5	5.9	5.7	5.7	5.9	6.9	7.0	7.1	6.6

Source: Computstat; SCCBD (2007).

innovation' to 'self-dominated innovation', from 'General Technology Ownership' to 'Core Technology Ownership' (*Wenhui* newspaper, 28 August 2006). All these efforts strengthened its innovation ability and raised its international status and competitiveness.

The development of exploitation and exploration strategies in DPOEs

The development of exploitation and exploration strategies at Stage 1 (1978–1992)

Institutional factors influencing the adoption of strategies

In the 14 years following the beginning of the reform in 1978, private firms were considered an inferior ownership form for ideological reasons. The overall political environment was antagonistic towards private business in the early years of reform (Dickson, 2003; Young, 1989). During this period, DPOEs were not significant. Few private firms had access to formal finance and only well-connected private firms could gain access to formal-sector finance (Haggard and Huang, 2008). Private entrepreneurs had to deal with hostility and social prejudice on the part of cadres and people in general who regarded them as dubious, ignoble or even despicable.

A large market for labour-intensive consumer goods was left unaddressed prior to the 1980s as a result of the heavy industry-oriented development strategy pursued by China (Lin et al., 1994). At the end of the 1980s, demand for durable consumer goods picked up. Although attention in China moved away from heavy industry toward the lighter type after the reforms of 1978, the growth mainly happened in durable consumer goods such as television sets and refrigerators. In the 1980s, private firms were smaller and it did not take much to start up a firm (Haggard and Huang, 2008). This gave the DPOEs a perfect opportunity to fill the gap (Wang and Yao, 2001).

The adoption of exploitation and exploration strategies of DPOEs

During the early phase of the reform in the first stage, Yankon and Tiantong[1] could not afford to exploit and explore simultaneously because the two firms had limited available resources. The manufacturers mainly competed on price to grab market share.

Resource exploitation

The uncertain regulatory environment for the DPOEs such as Yankon and Tiantong made for quite different ways to develop. The DPOEs were pushed to cost-based competition using cheap labour and abundant raw material. From the late 1970s to the early 1990s, the two cases mainly operated in the market gaps that larger SOEs or COEs had not explored (Interview 53). Yankon and Tiantong relied on their low-cost advantage and mass production modes, rather than product innovation, as their main competitive advantage. Compared with SOEs and COEs, the two DPOEs were highly profit-driven and adventurous, but with few financial resources. They tended to manufacture products with a simple structure and technology. For example, Tiantong specialized in mid- and low-end products. Raw materials accounted for 50 per cent of Tiantong's costs during its early development. The advanced technology was not expected to contribute greatly to its products during this period (Interview 49). Yankon was one of the first DPOEs in China to enter the energy-saving lighting sector and it developed a series of products of this kind (*New Finance*, 2004). According to the senior management of Yankon, unlike the business of SOEs working in that area, Yankon tried to avoid competition with high-end products (Interview 49). Yankon positioned its designs for medium- and low-end products during the early stage of reform (*PR Newswire*, 2005).

The development of exploitation and exploration strategies at Stage 2 (1993–2007)

Institutional factors influencing the DPOEs

The absolute number of DPOEs as a class expanded quickly after 1993. The central government promulgated legal codes to improve the business environment for DPOEs. In 1996, the legislation of the 'Law of Township and Village Owned Enterprises' provided equal legal status to collective and private TVEs (Wen, 2002). Private ownership was formally incorporated into the Chinese Constitution in 1999. In 2002, the Chinese government revised the constitution to recruit private entrepreneurs into the CCP (Dickson, 2007). These laws implied that the Chinese government provided equal legal status to private firms (Newell, 1999). For instance, the Ministry of Commerce and the Ministry of Science and Technology listed Yankon, Silan and Tiantong among the top 100 key exporting firms in 2005 (*www.most.gov.cn*). More than 40,000 DPOEs obtained the right to engage in external trade, which had been monopolized by SOEs and earlier by government-run foreign trade corporations (Ozaki, 2004). Local government officials also relied on DPOEs for developing local economies through employment and tax contributions.

The adoption of exploitation and exploration strategies of DPOEs

Resource exploration

In the late 1990s, the DPOEs gained a growing level of freedom to operate. They played an increasingly important role in the export of products. One of the outcomes of the low cost and low technology was that their products were easily replicated by competitors. To address this issue, the senior management of the three DPOEs developed an export-oriented strategy to gain overseas markets, giving their companies a competitive advantage over local rivals.

To overcome fierce competition with copycat products in the domestic market, Yankon thus envisioned a strategic shift in the company's focus from the Chinese market to foreign markets (Interview 43). Nearly half of Yankon's products were exported to Europe, North America, Japan, Korea, Taiwan, Hong Kong and Southeast Asia. Yankon's export sales surpassed US$100 million in 2005. This was the highest of any of China's lighting manufacturers (*Shangyu Daily*, 2006). Figure 4.2 indicates that Yankon's exports accounted for the lion's share in total sales revenue. One manager explained that the prices of Yankon's products were much lower than those of foreign countries. Because it possessed resources of cheap labour and rich raw materials, the unit price of Yankon's products was US$1 compared with US$5 for comparable products made by foreign countries (Interview 43).

The export share for both Tiantong and Silan also increased. For Tiantong, its main consumers were located in China, but more recently half of its buyers were from Taiwan, the USA and Japan. Fifty per cent of Tiantong's sales were in the domestic market while the other half were overseas exports, with a gross margin booked at 23.5 per cent in 2007 (Interview 49). Silan's exports climbed up from 27 per cent of the sales revenue in 2002 to over 50 per cent in 2007 (Interview 49).

Figure 4.2 Yankon's overseas and domestic sales revenue, 2001–2005

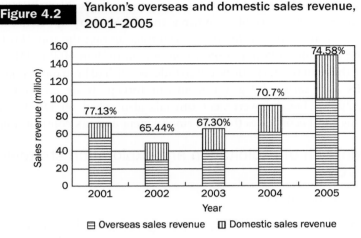

Source: Interview (33).

Note: The figure above each column represents the percentage of overseas sales revenue compared with the gross sales revenue of Yankon.

Although the managers of DPOEs acknowledged their advantages (such as the price of their products), they also noted the constraints that had prevented them from development, such as more difficult access to financial resources and lower social status (Interviews 33, 35, 36, 42, 47, 48). In order to compete and survive, they had had to acquire additional resources and capabilities. Listing on the market was employed to acquire the two kinds of resource (Interviews 37, 42, 49).

One of the biggest impediments to DPOEs was a lack of adequate financial resources for technological development or equipment improvement. Given their thirst for capital, the three firms sought to raise capital from the stock market. All the interviewed managers from the DPOE cases viewed listing on the stock exchange market as an effective way to achieve growth (Interviews 33, 35, 36, 37, 40, 42, 47, 49). According to an interviewee from Silan:

> The motives behind this strategic decision [of listing] included seeking external sources of finance and establishing a good social image in order to gain competitive advantages and achieve further growth. (Interview 20)

Listing on the stock market raised capital from the public and further helped the firms to obtain loans from the banks. Due to their relatively short history and their lack of reputation, the banks sometimes found it difficult to obtain financial data about these firms. One manager from an SOE expressed the difference between a SOE and DPOE in China in this way:

> Although the private sector is growing rapidly, the banks in China are still reluctant to lend money to private firms. If the loan made to SOEs is not repaid, the banks are probably only responsible for the economic mistake; if the loan made to the private firms is not repaid, the banks have not only made an economic mistake, but also a political mistake. (Interview 5)

Their listing gave them a social credibility they otherwise would not have. This was because one of the listing requirements was to submit the firm to more rigorous accounting and to provide better external reporting of accounts and performance. Since many private firms were reluctant to reveal their internal information, such firms might not want to be listed on the stock market unless they needed large amounts of long-term capital. The social standing of privately owned firms had improved greatly since the economic reforms in China. However, the fact that private entrepreneurs originate mainly from the lower class still negatively influences the social image of privately owned firms in China (Interview 37). Managers from the three firms generally agreed that they have better creditworthiness than unlisted DPOEs and that most of their loan applications had been successful (Interviews 35, 36, 39, 40, 44, 49). Reporting regulations for listed firms was helpful in making the DPOEs more attractive to potential external sources of finance for their continued growth. As the secretary to the board of Silan stated, 'it is relatively easier for a listed DPOE to get the loan from the banks' (Interview 38).

Senior managers of these firms agreed that in addition to raising funds and improving the market performance, listing on the stock exchange market also improved their image. Listing transformed a company into a public company, raised its social image or standing, and in turn strengthened its capability in dealing with the government (Interviews 36, 38, 44).

Resource exploitation

The lack of advanced technical and managerial knowledge available to such firms became an increasing constraint on growth. The top management of the firms realized that they could no longer remain competitive based solely on cost considerations and that they needed to be competitive based on quality (Interviews 35, 38, 45). Since the late 1990s, the firms had undertaken different journeys in the field of innovation and technology. In order to acquire the technology and managerial knowledge, the three cases had adopted different approaches to technological innovation: OEM and technological transfer. Compared to the other domestic players, Tiantong had advantages in its technology in addition to its scale. Tiantong focused more on importing equipment from advanced countries, such as state-of-the-art tunnel kilns with automatic nitrogen protection and eight lines of covering kilns from Germany and Japan, which improved the quality of products (Interviews 46, 49). Yankon focused more on the combination of in-house development and OEM (Interview 36), while Silan focused on the use of technology transfer (Interviews 41, 42). Their foreign collaboration reflected their strong commitment to technological modernization as well as their decision to select appropriate technology from different sources that promoted their ability to learn and control the technologies acquired.

Outside experts or managers have been recruited by private firms to enhance their competencies, although the original owner would need to surrender a degree of control (Child and Pleister, 2003). For the three cases here, recruitment of human talents have been among the most valuable of the acquired resources. They brought outside experts and competent managers to their companies in order to strengthen their internal financial reporting as well as to handle the increasing complexities that come with growth, diversification, and eventual internationalization (Interview 36, 42, 45). For example, in the mid-1990s, the Pan family of Tiantong employed an engineer from Shanghai with the annual salary of 50,000 yuan (about US$6,000), which was rare in family-owned enterprises in China when the average annual salary in the country was less than $450 at that time (Interview 45). By the end of 2007, Tiantong had over 160 professional staff and its own R&D centres. The manager of Yankon observed that the ranking position of the R&D staff was probably not high, but their salaries were higher than those of the senior managers (Interview 37). Silan designed an incentive mechanism for evaluating the technological achievements of the research staff (Interviews 41, 42). The award was divided into two kinds: one was the award based on the basic technological research; the other was based on the achievements of the technology. Between 0.1 and 0.2 per cent of sales revenue was put aside to be allocated as an award when the product reaches the market.

Results of the business strategies

The different business strategies that firms adopted evolved into different competitive positions. For enterprises that were unwilling or unable to change, the consequences were clear – a shrinking market share or decreasing sales growth rate. Table 4.3 indicates the domestic market share occupied by each of the cases respectively. Although fluctuating over the years, all the firms except Panda had a prominent share of their domestic market segment and ranked among the top five for market share from 1995 to 2007.

Table 4.4 describes the comparison of the sales growth rate of the eight cases with the average situation in the same sector. The comparison of growth rate in 2007 demonstrated that, except Panda, the other seven

Table 4.3　The firms' domestic market share, 1995–2007

	1995 (%)	2000 (%)	2003 (%)	2007 (%)	Ranking by market share (2007)
Hisense	2.9	10.9	11.1	16.2	1
Panda	9.7	3.7	3	5	N/A
Huadong	25	20	19.5	20	1
Haier	22	33.4	22	25.6	1
Chunlan	14.1	16.4	10.8	12	5
Yankon	3	7	10	15	2
Silan	N/A	5	17	20	1
Tiantong	14	14	30	35	1

Source: Chinese Household Electronics Report (various years).

Table 4.4　The growth rate of the main business revenue of firms, 2007

	Hisense	Panda	Huadong	Haier	Chunlan	Tiantong	Yankon	Silan
Growth rate (%)	25	−10.4	23.1	21	30.2	26.3	26.1	39.1
Average rate (%)	17.2	20	15	12	13.8	15	15	19.2

Source: Annual Reports (2008).

cases were leading domestic manufacturers in respect of their annual sales growth rate, which was markedly higher than the average growth rate in their respective sectors.

Conclusion

This chapter has integrated the constructs of March's exploitation–exploration framework with those from the strategy–environment–performance paradigm to propose a framework for analysing the process by which firms in a transition environment achieve a dynamic strategic fit (see Figure 4.3). In the Chinese context, Tan and Tan (2004) show empirically that the strategy–structure–performance relationship holds. However, the specific content of the strategy had coevolved with changes in the industry and the overall Chinese business environment. As shown in Figure 4.3, the implication is that competitive position is a matter of fit. Managers make strategic choices about what types of activities – exploitation or exploration – they adopt to allocate the resources so as to create a new fit. The managers need to make continuous adaptations until they realize the desired level of competitive position.

Regarding the research questions related to business strategies, this chapter has reached two conclusions. The institutional situation in the Chinese CE sector is the first to be discussed, as the firms have to face a

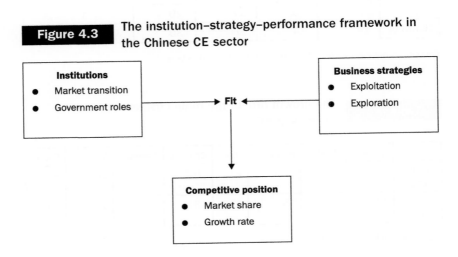

| Figure 4.3 | The institution–strategy–performance framework in the Chinese CE sector |

complex institutional environment and deal with institutional pressures when adopting business strategies.

The institutional situation in the Chinese consumer electronics sector

The findings suggest that institutional transformation exerts a strong influence on the adoption of business strategies. The transition towards a market economy in China brought huge institutional change to Chinese firms. The overall business strategies of the sample cases provided us with an opportunity to examine the external social environment in which they were developed. The emergence of a competitive market and the changing role of the government in economic activities would appear to be the two key factors, both of which reshaped the competitive landscape of the firms.

Market transition

The market transition the firms were facing in the Chinese CE sector can be described as the coexistence of intense market competition and lower entry barriers. With the deregulation of government policy in this sector, the entry barrier became lower. The number of new firms with different ownerships entering the sector far exceeded the central government's expectations. Lower entry barriers increased the pressure to improve efficiency. The increase in the number of firms and competitive intensity transformed the motive of the firms from accommodating the dictates of economic planners to meeting the demands of the market.

The market share and therefore the profits were more sensitive to production costs. In order to maintain a minimum market share for survival, managers must be motivated to work harder to adapt to market requirements. The motive for profit created by intense market competition changed the behaviours of all kinds of firms. Firms showed sensitivity towards changes in the market and shifted from fulfilling production quotas to making profits and thereby increasing their suitability for long-term development. The senior management of most of the companies in the study was positively influenced by market demands for efficiency, profitability and the achievement of high quality standards. With the transition of the market, most of the SOEs, COEs and DPOEs have adopted commercially oriented strategies unshackled from State demands that could often be inconsistent with the pursuit of profitability.

Role changes of the government

Competition disciplined local government agencies to promote and secure better ownership arrangements. The decentralization of central government allowed for the alignment of local government interests with local business, and local government was turned into the 'helping hands' of local business (Qian, 2002). Local governments and firm managers were given increasing autonomy over decision-making. Local government was actively involved in supporting local firms – regardless of their ownership – within its jurisdiction as long as the firms constituted the primary source of revenues for local government. This transformation of the role of government enabled firms to make decisions more freely with profit-making motives.

The 'helping hand' function of the local government function was shaped by transforming the criteria of evaluation towards the local government officials (Qian, 2002). Political promotion in China's new reform economy was tied closely to growth (Walder, 2003). Unlike in the planned economy, the superior economic performance during the transition period in a district in China was one of the significant criteria used in evaluating the achievements of local officials.

Along with the changing function of the local government, the relationship between firms and local government turned from unilateral dependence of the firms on government to bilateral dependence – from the sole heavy dependence of firms on government in the planned economy to the local government's dependence on firms as well, in the transition period in China. At the early stage of development, the SOEs and urban COEs operated in a business environment that lacked market factors. In some cases firms had to follow orders from the government, which provided money and protection. Under the current Chinese context of a market-oriented economy, local governments had incentives to provide more autonomy to firms to make market-based decisions that would improve competitive capability. These competitive firms in turn helped the local governments to fulfil their economic goals.

The evidence also revealed that although the government was deeply involved in bailing out large, failing firms, such cases were more the exception than the norm. Currently, only in extraordinary circumstances did State firms get such support from the government. For instance, part of Panda's business involved military-related electronics products, which were still monopolized by the central government. Most of the interviews demonstrated that the managers perceive government assistance as limited and unreliable in the long term. More often, government agencies acted as patrons or sponsors rather than as owners.

The fit between institutional environment and business strategies

The development of the firms in the area of business strategies experienced two major stages (Table 4.5). Environmental changes require changes in the firms – new resources and capabilities.

The analysis of the cases revealed the dynamic process of constraints and opportunities in their environment. In the reform process, the strategies and goals of the managers and their complement of resources and capabilities changed over time. Some firms made great achievements in terms of adapting to the market, thus making it possible to raise productivity and efficiency. Other firms, however, were slow or even unable to make adjustments to adapt to the changed market conditions and could not achieve a competitive position.

The choice between exploitation and exploration is important in a dynamic environment. The concepts of exploitation and exploration in the study build on the resource-based view of the firm, which assumes that a firm achieves a competitive advantage not only because it owns proprietary assets, but also because it possesses a superior ability to make good use of those assets (Barney, 1991; Conner, 1991; Penrose, 1959; Peteraf, 1993; Wernerfelt, 1984). More importantly, based on the transaction cost economics, this study focuses on how to exploit firm-specific resources and at the same time develop new ones to optimize fit with the external environment.

According to Masini et al. (2004), turbulent environments require continuous adaptation, which is often achieved at the expense of efficiency (less than optimal resource use). Accordingly, because maintaining or improving process efficiency was more difficult in such environments, those companies that manage to do so successfully were expected to experience higher benefits. The increase in resource exploration strategies was proportionally more beneficial for firms. While this study acknowledges the view of Masini et al. (2004), the findings in the study yielded different results: exploitation was a more efficient means to acquire resources needed at the first stage in the Chinese CE sector. When the firms entered the second stage, over-reliance on exploitation changed to the duality of exploitation and exploration (see Table 4.5).

In the first stage of Chinese reform from 1978 to 1992, although central planning persisted, the reform process incrementally improved incentives and expanded the scope of the market for resource allocation. The exploration strategy of firms within the CE sector during this period was

Table 4.5   The stages of firms' pursuit of business strategies

		Hisense	Huadong	Panda	Haier	Chunlan	Yankon	Tiantong	Silan
Stage 1 (1978–1992)	**Institutional factors**	Still government support for SOEs but the support was declining; increasing competition in domestic and broad market			Lack of protection from the government		Political environment was antagonistic towards private business; demand for durable consumer goods picked up		
	Business strategies	**Exploitation:** acquire knowledge through alliances from buying patents and licensing to establishing joint ventures **Exploration:** create independent distribution networks		**Exploitation:** heavily dependent on government **Exploration:** in-house development of technology	**Exploitation:** strategic alliances and 'buy in' **Exploration:** create distribution network		**Exploitation:** low cost and mass production **Exploration:** N/A		
Stage 2 (1993–2007)	**Institutional factors**	Government further relinquishes control over SOEs; increased market competition, globalization and technological innovation			More attention from the government owing to their success; operated with a high level of managerial independence		Equal legal backing to collective and private TVEs		
	Business strategies	**Exploitation:** further expanded distribution network; international partners and 'buying in' **Exploration:** technical centres; located research centres in advanced countries; R&D in quality enhancement		**Exploitation:** strategic alliances for advanced products, but without developing own brand **Exploration:** N/A	**Exploitation:** further expanded distribution network and marketing activities; acquisition; bought in technology; trained its own employees **Exploration:** brand building; building production parks overseas; establishing central research institute		**Exploitation:** brought in outside experts and hired competent managers; imported equipment; made strategic alliances for collaborative development **Exploration:** expansion of product market overseas; listing on the market		

mainly centred on sales and distribution, which grew rapidly as managers became more attuned to the workings of the market. Less reliance on exploration during the first stage may be explained in several different ways. First, the resources and capabilities of the firms in the first stage were not sufficient to pursue an exploratory strategy. Dierickx and Cool (1989) stress that the amount and level of a firm's resources and capabilities are the primary determinants of an exploration capacity. Firms with limited available resources were not able to afford to explore extensively. Second, firms that tried to pursue an exploration strategy faced higher risks. The cases here showed that leading firms in China often depended on the exploitation of advanced technology. With high technological turbulence and tight budget constraints, staying ahead of competitors for new technology was not an easy task. The huge cost and high risks involved in the development of radically new products were extreme. A firm's ability to obtain knowledge faster than its competitors was a key component of its competitive advantage at the first stage. Consequently, licensing and cooperation were often identified as proper strategies for acquiring technology. Third, the duration of time for measuring the impact of exploration is lengthy (March, 1991). It takes some years to achieve positive outcomes from exploration activities (Teece et al., 1997). Efforts in these cases were usually directed towards exploitation, since the outcome from innovation was hard to predict. Unlike exploration activities, exploitation in the form of licensing and forming alliances reduced the time horizon of research projects and allowed firms to acquire new skills and competencies at the same time, which strengthened their competitive position in the short term. Consequently, attempts at exploitation are a more appropriate strategy at the first stage.

During the second stage, the decentralization of economic authority from the central to local governments triggered competition. A supportive economic and social environment was crucial in providing incentives for technological learning (Xie and Wu, 2003). Without it, even formalized technological learning may be inefficient. The firms faced a real competitive environment in the domestic market. In order to survive in a product market and increase local revenue, local government provided the managers of firms with strong incentives for competition. To meet the new competitive requirements, firms faced the need to make changes in their business strategies, in particular to embrace aggressive learning strategies. Firms found it difficult to leverage their technological knowledge without controlling key complementary assets. Over-reliance on exploitation can be detrimental to the success of firms (Katz, 1997; Miller, 1990). Focusing too much on exploitation did not allow them to develop a stable, efficient manufacturing process. Firms accumulated and upgraded their distinctive resources and capabilities through the exploitation process at the first stage (Isobe et al., 2004), which in turn enhanced their innovative activities and investments at the second stage. Researchers in the management field have frequently argued that exploitation and exploration involve competing tensions within organizations (Levinthal and March, 1993; March, 1991). However, the results in this study contradict this perception. At the second stage, strategic fit resulted from the duality of strategies and the coexistence of extremes rather than by choosing one or another. The experience of these firms suggested that there may be different processes but that they are not necessarily antagonistic. Close inspection of the cases' strategies revealed that managers were simultaneously exploiting and exploring in the transition period to achieve a competitive position (Table 4.5), and the firms had to strike a balance between these two choices. They were only able to do so because they could judiciously allocate resources for exploitation and exploration activities in their pursuit of a dynamic strategic fit. In addition to the collaborative alliances to codevelop new technologies, the firms internalized R&D capabilities through a combination of acquisition and internal developments. The managers who undertook both activities that had a fit with the environment were successful in the sense of achieving a bigger market share and a higher sales growth rate.

The Panda case highlighted the misfit between the new environment and strategic choices. It was like a non-swimmer being suddenly plunged into the 'sea of the market' by the force of reform (Wu and Li, 2006). At the first stage, Panda did not realistically assess the opportunity costs and probable outcomes of huge R&D investments, which are more likely to be afforded by better-funded and better-endowed firms. At the second stage, Panda's managers still lacked the incentive mechanisms to reorient their operations toward consumer needs. They resorted to government support and help as an important strategy, when other firms continually increased their investment in R&D and innovation. The comparison between Panda and the other cases might provide lessons for latecomers in transition countries such as China.

Note

1. Since Silan did not exist until 1997, it is not included in the discussion in this stage.

5

Ownership structure and the characteristics of the board in the focal companies

Among the transition economies, the Chinese case is particularly intriguing. The Chinese economic reform began with decentralization rather than the development of a private ownership system, and with revitalization of State firms rather than private firms.

(Shaomin Li, Shuhe Li and Weiying Zhang)

Abstract: Although corporate governance has been a central issue in developed economies for some time, many findings from developed countries are not applicable in transition economies. Patterns of corporate governance and control differ significantly across countries because of national differences in the structure of ownership and the composition of corporate boards. This chapter aims to explore the character of corporate governance that has developed in China. The 'China model' is likely to embody a special role for the State, coupled with Chinese cultural aspects, while taking on some of the characteristics of the models in developed countries.

The chapter achieves this by empirically examining the characteristics and transformation of corporate governance in the sample firms. Using a multi-case study method, we compare the governance, ownership structures and the role of the board in managing firms. We find there is no single optimal form of ownership or board structure that produces a better performance. We also find that the better the performance of firms, the more advanced is corporate governance. We reject the view that corporate governance necessarily has an influence on performance. This chapter gives the reader an insight into the ongoing development and reform of ownership structures and the governing boards of firms in China.

Key words: ownership, board, China, case study.

Introduction

Corporate governance was a concept that was not high on the reform agenda in China until the late 1990s (Lu, 2002). Early enterprise reform focused on devolving decision-making to the managers of State or publicly owned firms, a process characterized as 'growing out' of the State-administered planned economy (Naughton, 1995). With the establishment of stock exchanges in the early 1990s, initially under the control of the local governments in Shanghai and Shenzhen, State firms increasingly began to reorganize their structure and offer shares to the public. In 1992 the newly established China Securities Regulatory Commission (CSRC) took control of public listing (Green, 2003), while no less a person than the then president Jiang Zemin proclaimed at the Communist Party congress in 1997 that the shareholding form of enterprise was appropriate for China. This chapter examines the characteristics of corporate governance of the sample firms, focusing on the ownership structure and characteristics of the governing boards. In an economy undergoing a process of system transformation, corporate governance is similarly in a state of dynamic flux. The eight case study firms have widely different ownership structures that allow a comparison of the corporate governance and the role of ownership structures and the board in the management of the firms.

This chapter comprises three main sections. The first examines the ownership structure of the firms. Three firms are State owned, two are collectively owned (a quasi-State owned firm) and three are privately owned. The next section investigates the organization and attributes of the board of directors and supervisory board. The last section explores the difference between corporate governance found among firms with different ownership types in China and compares these firms with corporate governance elsewhere.

The ownership structure of the firms

The ownership structure of SOEs

This section analyses the ownership structure of the three SOEs – Hisense, Huadong and Panda. We begin with a brief discussion of the pre-reform ownership structure of SOEs. A feature of most Chinese SOEs is that when they list on the stock market they do not list the entire company but only the best-performing units (Green, 2003; Qu, 2003). This section looks at the structure of ownership of both the parent holding company and the listed unit, and discusses the implications for governance.

The pre-reform ownership structure of SOEs

Until the early 1980s, the SOEs were stand-alone factories wholly owned, managed and operated by various levels of government (see Figure 5.1). Above the factory was a complex array of government agencies that administered every aspect of the factory. In effect, they were not companies in the Western sense, but rather a set of administrative production and distribution units subject to the direction of the Chinese State (Groves et al., 1994; Jefferson and Rawski, 1994; Naughton, 1995).

| Figure 5.1 | The pre-reform relationship between SOEs and Government |

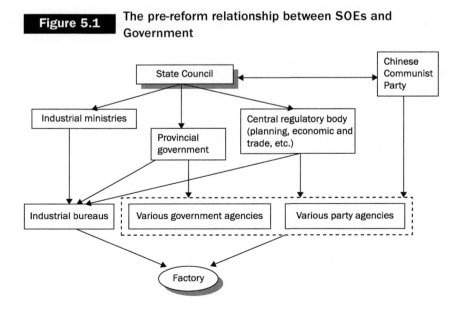

To control SOEs more effectively, the Chinese government developed a multi-tier system of control for exercising the State's ownership rights over the SOEs (Tenev et al., 2002). First, ownership rights were exercised at every level of government, from the central government to the provincial, municipal and township levels of government. Second, ownership rights were shared by a wide range of Party–State (government and Party) agencies. The Party–State agencies exercised the following rights: a) to appoint and monitor managers; b) to make key business decisions; c) to provide finance, including the entire investment fund, wages and salaries, and collective welfare expenditures.

Both control and ownership ultimately belonged to the State. In return, all profits of SOEs were turned over to the State. Government bureaucrats controlled the SOEs. The employees and managers had few incentives to maximize profit or efficiency under such conditions (Chow, 2002; Jefferson and Rawski, 1994; Naughton, 1995). Managers therefore did not have autonomy over production decisions, marketing and sales, and wage determination. The lack of autonomy resulted in an absence of incentives to improve the enterprises or to strive for efficiency and profitability.

In sum, the pre-reform (or traditional) SOEs were directly run by the central government and 'regulated and protected from both internal and international competition' (Huchet and Richet, 2002; Naughton, 1995). As governments gradually retreated from direct administrative control, most SOEs engaged in a two-phase restructuring to create a new company. Figure 5.2 outlines the process, which will be discussed in more detail below. After the restructuring, the original SOE remained mostly in the form of a parent holding company, while the new company was listed with a stock exchange (Tam, 1999; Tenev et al., 2002). In most cases, the newly established company would hold most if not all the profit-making assets, and the original SOE retains the unprofitable or less-profit-making assets (Lin and Zhu, 2001; Su and Jefferson, 2003). In return for the assets it injected, the original SOE received non-tradable shares in the listed company.

Restructuring ownership in the SOEs

During the first-phase restructuring, the three SOEs among our cases were corporatized (*qiyehua*), which was the term favoured for privatization in China to describe the spin-off and listing of parts of SOEs. Before an SOE could begin to restructure and list on the stock exchange it required approval from the local government, the State

Figure 5.2 The listing process for SOEs

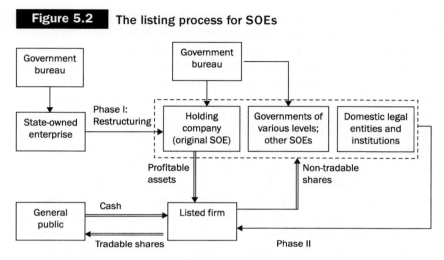

Source: Adapted from Liu and Sun (2005: 114).

Economic and Trade Commission, the State Commission of Economic Restructuring and the CSRC. Permission to list would stipulate that a quota of total shares should be issued. The main aim of the first-phase restructuring (see Figure 5.2) was to reduce the direct involvement of the government and transform the company into a legal entity with limited liability (Xu and Wang, 1997). This usually involved creating a holding company that held the assets of the SOE – both the listed and non-listed parts.

The firms then underwent a second restructuring (Figure 5.2). Part of the company listed during this phase, with a proportion of company shares, was offered for sale through the issue of an Initial Public Offer (IPO). The majority of the shares of the newly listed firm remained controlled by its parent SOE, governments of various levels, other SOEs and domestic legal entities. These shares were non-tradable. Only the publicly held A-type shares were tradable on the exchange, despite the State-tied shares comprising part of the total share capitalization of the firm. The purpose of the listing was to help facilitate the raising of capital, restructure ownership and invite private investors to contribute additional capital (Chen, 2005; Garnaut et al, 2005).

Hisense, Panda and Huadong issued their IPOs in 1997, 1996 and 1997 respectively (Shanghai Stock Exchange, *www.sse.com* and Shenzhen Stock Exchange, *www.szse.cn*). The original SOE became the parent holding company of the listed firm and received non-tradable State or legal person (LP) shares in the listed company. Hisense listed on the

Shanghai Stock Exchange and Huadong on the Shenzhen Stock Exchange. Panda listed on both the Shanghai and Hong Kong stock exchanges (*www.hkex.com.hk*). The Shanghai listing was A-shares denominated in RMB and only available to domestic Chinese investors and selected foreign institutional investors, while the Hong Kong issue was H-shares, denominated in Hong Kong dollars, that were traded in the Hong Kong market. As a result of the ownership restructuring, the holding companies became the largest shareholder of each of the listed firms. At the end of 2008, these State shares still accounted for 48.4 per cent, 27.7 per cent and 54.2 per cent of the total issue in Hisense, Huadong and Panda respectively (Figures 5.3, 5.4). The non-tradable feature of State-owned equity preserved State ownership in the listed firms. Accordingly, corporate control stayed with the State while private equity capital entered the firm.

As shown in Figures 5.3 and 5.4, all three companies developed a four-tier model of ownership after the restructuring. The first tier was government, either the central or the provincial government in our case study firms. The second tier included various State asset financial bureaus such as the State-owned Assets Supervision and Administration Commission (SASAC)[1] or other State agencies for management of State

---

Figure 5.3   **The ownership structure of Panda – Form 1**

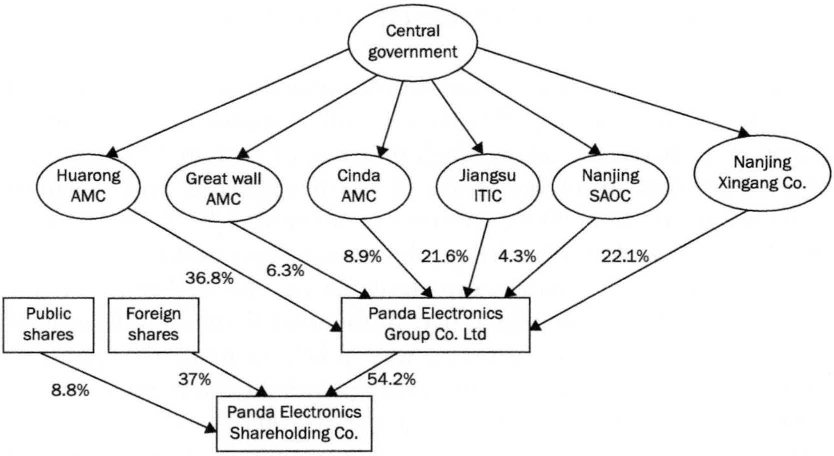

Notes: Arrows indicate control rights.
AMC refers to Asset Management Company.
ITIC refers to International Trust & Investment Corporation.
SAOC refers to State Asset Operation Company.
Source: Panda Annual Report (2008).

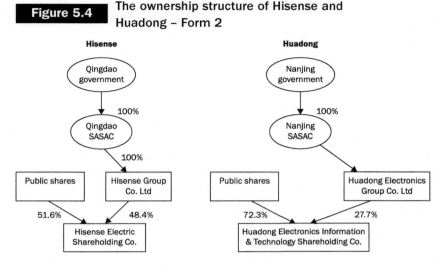

Figure 5.4 The ownership structure of Hisense and Huadong – Form 2

Notes: Arrows indicate control rights.
SASAC refers to State-owned Assets Supervision and Administration Commission.
Sources: Hisense Annual Report (2007); Huadong Annual Report (2007).

assets that were supposed to exercise shareholder rights in parent SOEs. The parent SOE resided in the third tier. The fourth tier was the listed entities of the SOEs. Yet the common four-tier form of ownership structure differed slightly between Panda on one hand and Hisense and Huadong on the other.

The ultimate owner of Panda was shared between the government at central and local level. Panda enjoyed significant preferential treatment from the government because of its military–strategic importance, and even more so since 1998 when the firm became mired in financial troubles. In 2005, in view of a contractual dispute between Panda and the Bank of China, 193 million shares of the listed firm held by holding parent company were frozen by the Higher People's Court of Jiangsu Province on 17 November for one year. On 14 November 2006, the freeze on the shares was extended for six months to 13 May 2007 (Interview 55). The government rescued Panda through a negotiated agreement with the bank instead of allowing Panda to be wound up. According to the agreement, the State-owned shares of Panda were acquired by six State-asset management companies[2] (see Figure 5.3).

After this forced restructuring, Panda's largest shareholder became the Huarong Asset Management Company (AMC), a central government-owned company that acquired 36.84 per cent of the shares. The remainder was distributed among five other State organizations: Nanjing Xingang Hi-Tech Co. Ltd (22.1 per cent), Jiangsu International Trust & Investment Corporation Ltd (ITIC) (21.6 per cent), Cinda Asset Management Company (8.9 per cent), the Great Wall State Asset Management Asset (6.3 per cent), and Nanjing State Asset Operation Co. Ltd (SAOC) (4.3 per cent). Huarong, Cinda and Great Wall were three of the four principal central government-owned AMCs.

Given that the role of the AMCs is to bail out the debts of State firms, if one of these four AMCs became a shareholder of a listed firm, it implied the firm was not performing well, operating at a loss or unable to repay its debts (Interviews 51, 56). To alleviate the debt burden of Panda, the three AMCs organized for some of Panda's debt to be disposed through a debt-for-equity swap (*South China Morning Post*, 13 May 2006).

The preferential treatment enjoyed by Panda was due to its special status. Not only was Panda one of the first national brands in China, the firm also played an important role in the national defence sector. According to the account of one manager at Panda:

> Panda is a comprehensive company not only in the area of consumer electronics, but in military communication. In this sense, the decrease of State shares in the company is not beneficial to its development. The development of military products in China cannot be totally dependent on non-State companies due to national security. (Interview 9)

Government assistance to the company in times of its financial distress enabled it to survive, but the over-dependence of Panda on government support had its drawbacks. It was a disincentive and sapped motivation in the managers. Panda's managers attributed their lack of competitiveness to inadequate protection from the government rather than to weak managerial initiative.

> I think there are weaknesses in the government policies. They do not guide, protect, and support the development of SOEs. The Chinese firms in the consumer electronics sector are not competitive enough in the global market at present. But the government forced us to join the world competition. The result is that we could not survive. (Interview 12)

Our informant above quite clearly hankered for the days of more orderly State administration of the economy, rather than the anarchy and competitiveness of the contemporary market.

Compared with the government's heavy involvement in the business of Panda, in which the proportion of State shares changed little, the degree of State ownership in Hisense and Huadong declined over time (Figure 5.5). The ownership structure of Hisense and Huadong differed from that of Panda. State control rights over Hisense and Huadong were devolved from the central government to provincial governments, and were administered via the local government SASAC in Qingdao and Nanjing respectively (Figure 5.4).

Decentralization provided incentives to the managers of these firms. Managers from the two firms said during interviews that under local government ownership the government became dependent on the performance of local firms for economic benefits, such as providing tax and creating jobs (Interviews 2, 6, 20). The dependence of local government on the performance of local firms gave each firm more bargaining power, which in turn created incentives for the managers to maximize the performance of the firms. One of the independent directors of Hisense elaborated as follows:

> Hisense is a new form of SOE. Its operational mechanism including acquisition of the resources and the assessment of the employees is market-based. Even if Hisense is State-owned, its owners and CEO are driven by performance incentives and objectives, such as profit maximization or market share maximization. (Interview 4)

Competitive SOEs like Hisense and Huadong placed their profit motive over other performance requirements. This is in essence comparable with

Figure 5.5 A comparison of state shares in the listed firms of Hisense, Huadong and Panda, 2000–2008

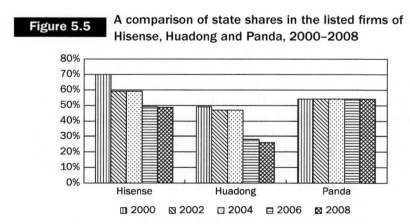

Sources: Annual reports of Hisense, Huadong and Panda (selected years).

the system of managerial autonomy that the government implemented in SOEs. The municipal government further delegated its power to the local financial bureau, the local SASAC. The local SASAC decided on the sale and purchase of State assets of the firm (OECD, 2004). The equity capital of Hisense and Huadong was held and looked after by independent State asset management entities, rather than under the guidance of a sole proprietor. A senior manager of Huadong observed:

> Actually the government realized the more successful the local business, the more stable is the local government. Thus the government was determined to separate itself from business and to make SOEs fully commercialized. The government has greatly reduced administrative interference in the firm's operation. We can almost make all major business decisions except [about] the sale of the assets of the firm. (Interview 18)

The ownership structure of COEs

The control rights of Haier and Chunlan as a COE technically resided with the workers or the local community, though in practice both companies were formally under local government control. The employees of COEs in practice had no claim on the residual profits of the firms. As economic liberalization allowed Haier and Chunlan to grow into larger entities, the laws governing their management and operation failed to keep pace, creating much ambiguity about their ownership.

The pre-reform ownership structure of COEs

The development of Haier and Chunlan illustrates the ambiguous ownership structure found among COEs. The two firms evolved from small manufacturing plants into large brand-name consumer electronics companies. Their expansion allowed them to acquire ailing SOEs, which resulted in their ownership structure becoming increasingly unclear. The Chinese government stipulated that the assets accumulated through investments financed from budgetary grants of the government or State banks loans are defined as government-owned (Liu and Sun, 2005). During the acquisition of ailing SOEs, Haier and Chunlan enjoyed special municipal government support, which included government grants and low-interest loans from State banks. In such circumstances, the property formed during the acquisition became State-owned assets.

Haier also accumulated company-owned assets separate from acquired State assets during the development. However, it was difficult to isolate the company-owned assets from all the State-owned assets. As a result, the interweaving of the State-owned property within Haier and Chunlan during the acquisition of SOEs and the company-owned property blurred their ownership status (*Operation* newspaper, 11 April 2004). The assets accumulated by the company itself were poorly protected by formal contracts and regulations.

The ownership structure of Haier and Chunlan described in Figure 5.6 suggests that the government was not involved directly in the company's ownership, but was at arm's length, as shown by the dotted lines.

Nevertheless, both companies in practice were influenced by government policies aimed at SOEs. For instance, in 2003, the local SASAC announced the list of 27 companies under its control, including the Haier Group among them (*Financial Times*, 15 December 2004). Since the primary focus of SASAC is to deal with State assets of SOEs, Haier's inclusion in the list seemingly classified it as a SOE. Neither senior management of Haier nor the heads from various levels of SASAC would disclose the details of why Haier was subject to the SASAC (*First Financial Daily*, 1 December 2004). Chunlan faced similar problems.

Figure 5.6   **The relationship between listed firms and ultimate controller in COEs**

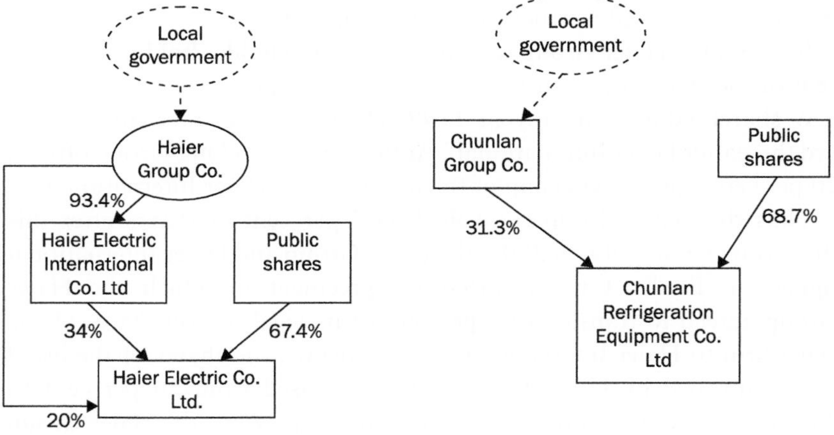

Note: Arrows indicate control rights.
Sources: Haier Annual Report (2007); Chunlan Annual Report (2007).

Although Chunlan had always been a COE, the National Bureau of Statistics classified its ownership as a 'partnership between State and the Collective' (*Shanghai Securities* newspaper, 2 November 2000). The unclear assignment of property rights motivated Haier and Chunlan to undertake different forms of restructuring to clarify their ownership structure.

Restructuring ownership in COEs

Uncertainty about ownership of both firms had disincentive effects on management. The top management felt a sense of insecurity about their personal interests in the company (*China Economic Times*, 13 August 2002). In the words of Zhang Ruimin, the CEO of Haier Group:

> It is impossible for me not to consider at all my self-interest in the company. No matter how successful the company is, if I left it one day, nothing would belong to me. (*China Economic Times* (in Chinese), 2002)

Transforming the relationship between the Haier Group and the local government was hard to achieve owing to the government's fear about the erosion of State assets. The increasing concern for their self-interest encouraged the senior management of Haier to find a solution through restructuring the firm.

Unlike the restructuring of SOEs, the Haier holding company, the Haier Group, ceased to be the largest shareholder of its listed firm during the management-led restructuring (see Figure 5.6).

In 1997, the Haier Group was the largest shareholder holding 54.7 per cent of the shares of the listed. The restructuring saw the founding of a new shareholding company in 1999, Haier Electric International Co. Ltd (hereafter Haier International). At this stage, the Haier Group offered 20 per cent equity ownership of the listed firm to Haier International.

Thus the Haier Group controlled 34.7 per cent of shares after this share transfer in 1999. In 2001, the Haier Group and Haier International signed an Equity Capital Injection Agreement in which the Haier Group agreed to transfer 14.7 per cent of its legal person shares of the listed firm to Haier International. Consequently, the shares of the listed firm controlled by Haier International increased from 20 per cent to 34 per cent and it became the largest shareholder, while the Haier Group became the second largest shareholder with 20 per cent of the shares (Interview 25).

Unlike the ownership structure of the Haier Group in which the State-owned and the company-owned assets were intermixed, the company-owned assets, including employee shares in Haier International, were clearly defined (*China Economic Times*, 13 August 2002). The new structure separated employee shares from State-owned assets in the company. Senior management therefore considered that these shares would preserve their interests in the company. In this sense, Haier International became a special-purpose vehicle to protect the benefits of management and employees. Haier expected to avoid the problem of asset intermixing by transferring the company-owned assets to another company and thus clarifying the ownership rights of individuals and legal persons (Interview 25).

Restructuring one of the entities of the company instead of the entire company provided an indirect solution to clarifying the ownership of the Haier Group. With the approval of the government, this achieved the goal of the company but avoided direct conflict with government policy. In the words of Haier CEO Zhang Ruimin, 'We work in a mixed economy. We have to have three eyes: one on the market, one on the workers, and one on the government' (Joseph, 2006).

In contrast, Chunlan sought a different way of transforming its ownership structure. To clarify the ambiguity of ownership and enhance employees' motivation, Chunlan tried to restructure the company's ownership structure using an employee stock option plan, which would grant to specified employees of the company an option to buy company shares. The employee stock option, mooted by Chunlan in 2000, was designed to align incentives for managers and employees with the performance of the company. According to an executive director:

> We are becoming increasingly aware that a good connection of the interests between senior management and the company can help bring about the sustainable development of the company. The ambiguous ownership of Chunlan quashes the employees' motivation and the company's vitality and consequentially lowers investors' confidence in us. We consider employee stock option is an effective way to reform the ownership of the firm because the employees are allowed to hold an adequate amount of the company stock and presumably will become more committed to their jobs. (Interview 30)

The company allotted a US$122 million share option, amounting to 25 per cent of the company's existing share capital (about US$122 million) to its senior managers and selected staff (*Shanghai Securities* newspaper, 2 November 2000). Employees who were to receive the planned share option would have eight years to exercise their rights under the scheme. Eligible persons included senior managers, directors of the board, research and development staff, and senior sales people.

The main provisions of the scheme were as follows (Interviews 27, 29):

- The selection of employees would be based on a minimum period of service. Employees were entitled to buy company shares up to a value equivalent to their annual income.

- A low-interest bank loan was to be provided to employees who had insufficient financial resources but who wished to participate in the plan. The loan was to be repaid within eight years.

- The employee share options were to be tradable on the stock exchange.

Within one month of the announcement of the plan, over 90 per cent of employees had signed up. According to the interviewees, the employee option plan was very successful in terms of enhancing the commitment of employees at the early stage of its implementation (Interviews 27, 29). However, the reform initiative faced difficulties. 'The restructuring of ownership is a sensitive topic in China if it touches upon State-owned assets,' an informant recalled (Interview 25). 'The initial investment was made by the State . . . [so] The reform must ensure that the companies retain the majority interest without irritating the government.'

The plan for Chunlan sparked a backlash from the central government. In the absence of any laws and regulations related to the transfer of State-owned shares to employees, this type of transfer was not allowed in China (Wei, 2002). For Chunlan it was difficult to identify whether the 25 per cent to be offered as share options included State-owned assets. The government feared that Chunlan would in effect transfer State shares to private individuals. Up to the end of 2007, Chunlan had been trying to work out a plan to satisfy all shareholders.

The ownership structure of DPOEs

The DPOEs case study firms – Yankon, Silan and Tiantong – each has a different ownership structure. Yankon and Silan are organized in the form of a holding company-controlled listed firm, whereas Tiantong is a listed firm controlled by a natural person.

The pre-reform ownership structure of the DPOEs

Yankon and Tiantong were initially family-owned enterprises, but were registered as township and village enterprises (TVEs), part of the rural collective-owned sector, while Silan was registered on its establishment as a privately owned firm in the form of a limited liability company.

The different forms adopted reflect the different institutional environment during their early development. Yankon and Tiantong were established in 1975 and 1984, a period when private ownership was not allowed. Yankon began as commune brigade industry, which became a TVE after the dismantling of communes in the early 1980s. Tiantong was registered as a TVE on its establishment. Adopting a collectively owned form was a political expedient and the only means to survive because of political discrimination against the private sector. This strategy was known as wearing a 'red hat' (Dickson, 2003). For example, when Tiantong was established in 1984, the founder made an arrangement with the township government whereby the firm would be officially owned by the authority and subcontracted back to the founder. According to the contract, the township government owned 36.3 per cent of the shares and was the largest shareholder of the firm (*China Securities* newspaper, 27 December 2006). The employees of the company, including family members, held only 9.3 per cent of the shares. The remaining 54.4 per cent of the shares, the collectively owned share of the company, was actually controlled by the township government indirectly (*China Securities* newspaper, 27 December 2006). The local authority would provide assistance in acquiring premises, licences and electricity and by dealing with the tax bureau on the firm's behalf. The 'red hat' was needed for political protection, which allowed Yankon and Tiantong to grow successfully at an early stage of development.

When Silan was established in 1997, the legal status of the private firms was more securely established in China as a result of the passing of the Law of Township and Village Owned Enterprises (1996). The growing level of freedom to operate had led to the rise of many private firms without the need to use the disguise of TVEs (Chen, 2007). Some TVEs also changed to share ownership or share listed companies. As a consequence of this, unlike the private firms that initially used a TVE form to prosper, Silan was first registered as a privately owned enterprise in the form of a limited liability company, in which the shares were divided evenly among the seven owners (Interview 40).

Restructuring ownership in the DPOEs

As the three DPOEs grew, they were handicapped by their lack of financial resources, typical of small and medium-sized enterprises in China and indeed elsewhere. To raise capital they went to a public listing. To do so they were required to improve their corporate governance, both to comply with stock exchange listing rules and to make the company more attractive to investors.

As shown in Table 5.1, with the IPO listing on the stock exchange, the proprietors of DPOEs retained significant control of the firms' assets.

Table 5.1   The relation-based positions in DPOEs, 2006

	Family members	Position	Percentage of shares (%)
Yankon	Chen Senjie	Chairman	6.88
	Chen Wei (son)	Legal person of Zhejiang Zhenli Technology Science Co. Ltd	4.63
	Chen Ying (daughter)	Owner of Zhejiang Zhenli Technology Science Co. Ltd	4.83
	Other family members	Owners of the Yankon Group	11.46
Silan	Cheng Xiangdong	Chairman	4.78
	Fan Weihong (friend)	Deputy chairman	4.64
	Zheng Shaobo (friend)	Deputy chairman and general manager	4.64
	Jiang Zhongyong (friend)	Board member	4.64
	Luo Huabing (friend)	Board member	4.64
	Song Weiquan (friend)	Chairman of the supervisory board	2.05
	Chen Guohua (friend)	Supervisory board member	2.05
Tiantong	Pan Guangtong	Chairman	11.03
	Pan Jianqing (son)	Deputy chairman and CEO	10.45
	Pan Meijuan (daughter)	Deputy general manager of sales department in a Shenzhen subsidiary	1.35
	Pan Jianzhong (son)	Deputy general manager of procurement department	1.20
	Du Haili (daughter-in-law)	Deputy head of general manager's office	0.89

Sources: Annual Reports of the sample firms (2006).

These firms had many characteristics similar to Chinese family businesses that had listed in Hong Kong and elsewhere (Chen, 2001). The proprietor founders controlled the company through a set of cross holdings and family members or close friends, despite the greatest part of ownership being in the hands of the investing public. Two types of ownership structure emerged as a result of restructuring in China.

In the first type, the family or the founders of the firm did not control the listed firm directly but preferred to control it through a holding company. Yankon and Silan were examples of this ownership structure, which reflected external financial constraints (Fan et al., 2005), as DPOEs were disadvantaged in their access to external funds in China. A senior manager and executive director of Yankon observed:

> China's State banks may hesitate to provide needed finance to private entrepreneurs, for they are newcomers and hence lack credibility. The State banks suspect whether the records presented by private firms truly reflect their financial position. In addition, banks may be willing to sacrifice profits in order to seek political, ideological or personal goals rather than the profits. If a privately owned firm defaults, State banks will bear greater responsibility than in the case of SOEs. (Interview 36)

This ownership structure was related to the external financing constraints of the DPOEs. Due to the difficulty of obtaining a formal loan, an entrepreneur's alternative was to create internal financial markets to allow cross-subsidization of funds between affiliated firms (Fan et al., 2005). Figures 5.7 and 5.8 show that Yankon and Silan controlled their listed firm with less than an absolute majority of equity ownership in the hands of the family or the founders. The ownership structure of Yankon shows that family members controlled the listed firm Zhejiang Yankon Group Co. Ltd through different holding companies including Zhejiang Zhenli Technology Science Co. Ltd, Shiji Yanko Group Co. Ltd and Hangzhou Yi'an Investment Co. Ltd (see Figure 5.7). The family had 50.1 per cent of control rights with only 10 per cent ownership. This was similarly the case with Silan. The seven ultimate owners of the listed firm had 65 per cent of control rights with 42.2 per cent ownership (see Figure 5.8). Control of the companies' assets could be obtained with low-ownership levels through the establishment of this pyramidal structure.

The second type of ownership structure occurs when the listed firm is controlled directly by individual persons or natural persons. Prior to

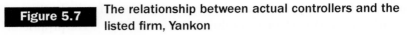

Figure 5.7 The relationship between actual controllers and the listed firm, Yankon

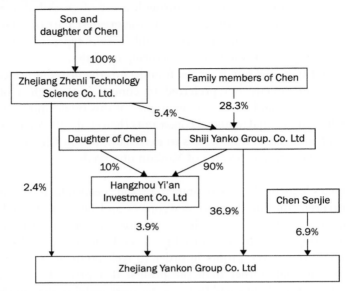

Notes: Arrows indicate control rights.
Ownership of the family: 2.4% + 5.4% × 90% × 3.9% + 10% × 3.9% + 28.3% × 90% × 3.9% + 28.3% × 36.9% + 6.9% = 10%.
Actual control rights: 2.4% + 3.9% + 36.9% + 6.9% = 50.1%.
Source: Yankon Annual Report (2007).

1998, Chinese laws prohibited natural persons from directly holding more than 0.5 per cent of total shares outstanding for any listed company. The Security Law of China has extended natural persons equal standing to legal persons and other social entities since 1998 (Chinese Securities Law, 1998) and apparently created incentives for them to register more transparently. Tiantong was the first natural person-controlled company in 2001 in China to be listed (*China Securities* newspaper, 27 December 2006). The father and son owned 21.2 per cent of the issued shares (Figure 5.9).

As one senior manager from the company commented:

> Tiantong's ownership structure allows a high consistency of the interest between the controller and the public shareholders. In addition, a natural person as the dominant shareholder pays more attention to the firm's social reputation, which in turn makes him more responsible to the company' business. (Interview 43)

Figure 5.8 The ownership structure of Silan

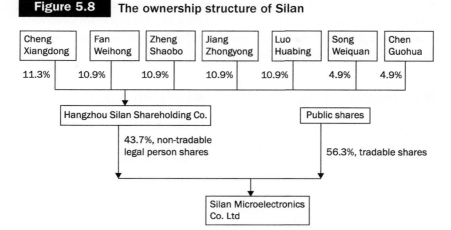

Notes: Arrows indicate control rights.
Ownership of the seven people: (11.3% + 10.9% × 4 + 4.9% × 2) × 56.3% + (11.3% + 10.9% × 4 + 4.9% × 2) × 8.7% = 42.2%.
Control rights of the seven people: 56.3% + 8.7% = 65%.
Source: Silan Annual Report (2007).

Figure 5.9 The relationship between actual controllers and the listed firm, Tiantong

Source: Tiantong Annual Report (2007).
Note: * The shares held by Pan Guangdong and Pan Jianqing are sponsor shares, which are also called sponsor legal person shares (*faqiren farengu*), a special form of LP shares that are not allowed to be traded.

A comparison of ownership structure in the cases at different stages

The restructuring of ownership over the past three decades is crucial to understanding the current situation of the firms in China and their future development (Fan, 2000). Prior to the reform, government at all levels played a direct role in firms, and ownership was vested in the State or collective. As Chinese economic reforms brought a change from a planned economy to a market economy, firms rapidly diversified ownership and the role of the State steadily diminished.

The creation of various share ownership types

Ownership reform is a central process in the development of Chinese enterprise reform. At the time of conversion to a shareholding enterprise, new forms of share ownership were created. Private capital appeared in the SOEs and COEs. Similarly, State capital was also injected into the DPOEs as in the example of the stake Shanghai Baogang took in Tiantong. However, as Table 5.2 shows, the legal person shares and tradable A-shares were the most prominent (or tradable).

Except Panda, the largest shareholders of the other seven case firms are legal persons. However, the ultimate owner or the identity of the legal person shares is qualitatively different. For DPOEs they are private institutions or individuals unrelated to the State, whereas for SOEs and COEs, the legal person owners are ultimately the central or local government, or another State-owned enterprise. The legal person shares in SOEs and COEs represent an indirect degree of State control over these firms. The official shareholding classification in China conceals the ultimate ownership. The absence of State shares shown in a company's annual report does not necessarily indicate the absence of ultimate State control.

Control via the parent holding company

Apart from Tiantong, which was directly controlled by a father, son and two major corporate shareholders, our other seven listed firms were controlled through parent holding companies. This practice is widespread in China. The purposes of the ownership structure differed among the firms. For the SOEs and COEs, the primary reasons were twofold. First, the existence of institutional constraint on firms arising from the central government's fears about the potential loss of State assets, and second, local government's desired to capture economic rents from these firms while allowing decentralization of managerial control rights. This process was consistent with the view that partial listing allowed local governments to decentralize the decision rights to firm management without selling off their ownership (Fan et al., 2005). Since local governments were constrained by Chinese laws that prohibit the transfer of State ownership, partially listing a firm was an effective means to decentralize the decision rights without loss of ownership and to motivate the firm's managers. For DPOEs, the holding company became the control vehicle of the family or firm founders, which enabled them to retain control over the listed entity but to attract outside investment to meet their financial needs.

Changes in ownership concentration

Non-freely tradable shares comprised two kinds: State-owned and legal person owned shares. From Table 5.2, we see that during the initial years of the firms' listing, the State and the legal person made up on average nearly 58 per cent of all the shares of listed firms. In Hisense and Chunlan, the largest shareholder actually accounted for more than 70 per cent of the total shares. The fact that State shares and legal person shares were not traded on the securities market means that 50 per cent or more of the issued share capital was excluded from market. Only about 26 per cent belonged to individual shares. With control power vested in non-traded State and legal person holdings, questions arose about the extent to which listing might subject the firm to market discipline, which was one of the aims of the listing.

By 2007, the ownership concentration compared with the initial year of listing had changed considerably in most cases. The concentration of ownership was reduced and more individual shares or A-shares were traded on the market. In contrast, State ownership in Panda remained unchanged since its first listing (see Table 5.2).

Characteristics of the boards of firms

The characteristics of the boards along with the composition of ownership are two of the biggest factors affecting corporate governance in China (Feinerman, 2007). Organization and conduct of the board are crucial to understanding the governance mechanism of the firms discussed in this study.

Chinese Company Law requires all listed companies to adopt a two-tier board structure, with a board of directors (BoD) and a supervisory board (SB). The BoD is the main decision-making board governing the

Table 5.2	A comparison of shares in case firms at different stages (%)

		State	LP	Public shares	Largest share type
First year the firms listed	Hisense (1997)		74.1	25.93	74.1 (LP)
	Huadong (1997)	–	57.6	42.4	57.6 (LP)
	Panda (1996)	54.2		45.8	54.2 (STATE)
	Haier (1993)	–	65.2	34.8	65.2 (LP)
	Chunlan (1993)		75.0	25	75.0 (LP)
	Yankon (2000)		52.6	47.4	52.6 (LP)
	Silan (2003)	–	74.3	25.7	74.3 (LP)
	Tiantong (2001)		29.4	70.6	29.4 (LP)
2007	Hisense		48.4	51.6	48.4 (LP)
	Huadong		27.7	72.3	27.7 (LP)
	Panda	54.2		45.8	54.2 (STATE)
	Haier		43.5	56.5	43.5 (LP)
	Chunlan		31.3	68.7	31.3 (LP)
	Yankon		45.0	55.0	45 (LP)
	Silan		43.7	56.3	43.7 (LP)
	Tiantong	12.3	37.7	50.0	37.7 (LP)

Sources: Annual reports of case study firms (various years).

firm while the SB is formally an independent board that monitors the actions of the executive management and the board of directors. China's corporate governance system is unique, borrowing elements from both the Anglo-American unitary board and the German two-tier board models (Dahya et al., 2003).

During the past decade or more there had been much development of corporate boards. Many initiates stemmed from recognition of the need for stronger corporate governance that was accepted in a 1999 resolution at the CCP's Fourth Plenum meeting of the Fifteenth National Congress (Lu, 2002). But the transitional nature of the Chinese economy created difficulties that inhibit the effectiveness of company boards. This section will discuss the characteristics of the boards from four perspectives: the decision-making powers of boards, the supervisory board, independent directors and the affiliated directors.

Decision-making powers of boards

Efforts to promote the independence of boards' decision-making had been a recurring theme since the 1990s. In the classic literature on company boards, their role is seen to be one of monitoring management and overseeing the reward, punishment or replacement of the managers (Mace 1986; Patton and Baker 1987). Granting more power to the BoD for independent decision-making in China could facilitate their restructuring to better meet competitive pressure in the market.

Previously in the SOEs and COEs, the complex management structure and the dominant presence of Party cadres remained a challenge for enterprise reform. Unsurprisingly, the State as majority shareholder typically intervened in any major restructuring of a company. A director from Panda said:

> The decision-making procedure was quite complicated before. There were two parallel-ranking groups of top decision makers: the three-new and three-old. The new referred to the board of directors, the supervision committee, and the shareholders committee; while the old referred to the Chinese Communist Party Committee, the employee representative Committee and the union. The biggest challenge was how to ensure the independence of the board of directors, and allow the board to have the biggest say in business decisions. (Interview 10)

A similar view of the past was recalled by a non-executive director of Huadong, who noted that the senior managers in SOEs were previously all nominees of the State.

> The board couldn't make decisions on top management personnel hiring matters, their job evaluation and remuneration package – it had no real power at all. The candidates for board membership of the listed firms were first elected by the holding parent company and the shareholders meeting endoresed the final board members. (Interview 17)

In his view, 'the biggest setback' for effective board governance of the firm in the past was the presence of the company Party Committee, whose powers overlapped with those of the board.

> As the Chinese saying goes – you can't keep two big tigers in one cage. (Interview 17)

The Party supposedly provided only political guidance, while the board had formal power to make decisions, but ultimately effective decision-making was possible only with the agreement of both, he added.

The formal independence of the boards among our case study firms appears to have increased in recent years. Compared with the previous practice that the Party Organization Department would appoint both the board and the management team, there is evidence the boards in SOEs and COEs could hire and fire management in accordance with their corporate charter, though the SASAC must be informed (Kang and Cheng, 2009). The boards of directors gained jurisdiction over hiring matters concerning top management personnel, and the government seemingly sought to draw a boundary between the board and the Party Committee in other business decisions. For example, Haier has made important strategic decisions with little reference to the Party Committee or the local government for a decade or more. In 1999, without reference to Party-state organs Haier decided to establish a refrigerator factory in the USA because of the high import tariffs and the following year it set up a US$40 million design centre in Los Angeles and a manufacturing facility in Camden, South Carolina (Interview 25). Reduced involvement of the State and Party in business and commercial decisions has also occurred at Hisense:

> The governments now hardly intervene in the business of the company. The board has much more decision-making power nowadays. Although the most important decisions require consultation with the State-authorized organizations . . . the local government would accept the recommendation of the board as long as it thinks they are reasonable and can improve the local economy. (Interview 3)

However, the situation at Panda was again at odds with that of the other case study SOEs. The views of the interviewee from Panda contrasted sharply:

> To be honest, it is very difficult to deal with the State organizations. They would like to intervene in every aspect of the company. All strategic decisions, even the criteria for performance, are decided by the government. (Interview 13)

In the DPOEs' early stage of development, decision-making power was highly centralized in the hands of an entrepreneur, similar to the

Chinese family business model typically found outside China (Wong, 1985). The patriarchal founder of the firm had huge flexibility to act. In theory, he could run the business as he saw fit without consultation with other family members. The CEO of Tiantong, Mr Pan Jianqing, attributed his early success to a firm structure that put 'the control of decision-making in the hands of the family' (cited in Cheng and Ma, 2006). Despite the board of directors' and the shareholders' meeting, the founders of private firms played a dominant role in making decisions. But progressively the proprietors of DPOEs introduced into the board more outside advisors who had a degree of autonomy that allowed them to influence the firm (Interview 37). According to one of the founders of Tiantong:

> The most striking feature of many private firms in China is that the firms tend to concentrate the decision-making power in the hands of a single man, in most cases the father of the family. In Tiantong, however, it is common in the board meeting for directors to have a heated debate and reject the proposals of the executive management. (Interview 44)

For example, Shanghai Tianying Investment & Development Company Ltd is one of the most important investors of Tiantong. When Tianying was newly established in early 2002, the management of Tiantong proposed to cooperate with Tianying. However, the board felt there were too many risks because little was known about Tiantong. The board asked management to reinvestigate the background and development potential of Tianying, including its capital structure and market prospects, before the proposal was finally accepted several months later (Interview 44).

A Yankon interviewee acknowledged that 'many top executives are family members of the founder or his friends' and 'few people dare to doubt the father-entrepreneurs' decisions'. Nevertheless, outsiders were brought in:

> . . . the company has to search for talented professionals for the board in the face of the pressures to improve the corporate governance and increase the profits. The reward for the professionals is quite good. First, their remuneration package is quite good. Second, their tenure is long, maybe a decade or longer, as long as they are trusted by the family. (Interview 37)

Private firms in China, despite the role of the family, found that competitive market pressures compelled them to reach beyond the family and close circle of associates to attract expertise for senior management and their company boards.

The supervisory board

China has adopted a dual board structure. The Chinese Company Law stipulated that all joint stock companies were required to establish a BoD and a SB whether they were listed or not. A typical supervisory board had three members, but could vary between two and five members. In addition, the 'Tentative Regulations for Supervisory Boards of SOEs' (2000) required members to comprise shareholders' representatives, employee representatives and relevant experts. However, there are specific regulations on how to implement the SB's duties effectively. Most SB members have affiliations with the State; there is no professional organization of SB members that is committed to setting the standards against which they must act; there are no provisions concerning rules of procedure, rules of voting, or rules of proposing and holding meetings of the SB. These shortcomings render the SB weak and ineffective as a monitoring mechanism. Three categories of problems facing the SB are outlined below.

A lack of clear duties

Although the Company Law in China defines the role of the SB as that of a monitoring mechanism, its functions and tasks in general are unclear and lack transparency. Some interview respondents seemed uncertain of the SB's function and confusion about the role of supervisory directors (Interviews 11, 31, 34). Since the existing laws and regulations were silent on crucial questions about the relationship between the board of supervisors and independent directors such the as allocation of powers and liabilities, their functions overlapped, especially in the areas of overseeing the directors and examining the firm's financial affairs. For example, a financial officer of Tiantong stated:

> The supervisory board generally has quite limited duties in the company. Sometimes the role and functions of SB board members overlap with those of the independent directors because both serve as monitors. It seems to me supervisory directors lack specific guidance as to how to carry out their duties effectively. (Interview 48)

This was accentuated by the SB members' lack of the technical knowledge required to monitor the directors and executives. Among the eight supervisors who participated in the study, only two had relevant specialist knowledge. The others included the company's Party Committee secretary and the head of the labour union of the parent holding company. Their lack of business knowledge and expertise constrained the supervisory members in their ability to monitor the company's compliance with laws, regulations and rules, to examine the financial statements of the firms, to review business performance and to evaluate the CEO.

A lack of power

In China, personal relationships (*guanxi*) mattered hugely in business, political and social activities. The prevalence of strong social connections meant that the supervisory board would struggle to achieve independence. The entrepreneurs and managers preferred to look for supervisors from among those they knew would support their proposals, and who would resign in extreme cases rather than oppose those who selected them to join the supervisory board. A supervisory member said: 'As a member of the same [firm] group, we have to protect its interest and do everything according to "unified standards" [stipulated by the firm]' (Interview 43).

The monitoring role played by the SB in the listed firms of SOEs and COEs in China, however, cannot but be influenced by the Party–State. The SOE listed firms Hisense and Huadong had three supervisory members and Panda had five, while the COE Haier had three and Chunlan just two. At least one SB board member was from the Party Committee of the parent holding companies. Huadong was an extreme example of all supervisory members being from the Party organization of the parent company or from the Party organization of a subsidiary of the parent (Annual Report of Huadong, 2007). In this case, the SB is nothing more than another arm of the State, able to exercise control over the listed firm. The interview data exposed an inevitable inertia in relation to the influence of the government on the SB. A supervisory board member revealed:

> Most supervisors are Party officials. Key individuals are selected by the government. It is difficult for the supervisory board to avoid influence by the party. Sometimes I have a kind of feeling that the boards of supervisors in listed companies exist like a non-functioning empty shell. (Interview 15)

In DPOEs the chairman of the board was typically the founder, as shown in Yankon, Silan and Tiantong (Annual Reports, various years). Although sometimes not the largest shareholder of a DPOE, the chairman appeared to have absolute control over the board of directors, shareholder meetings and the supervisory board. For example, the SB of Silan consisted of three members, one principal and two supervisors. The principal member of the SB and another supervising member not only were two of the founders of the firm, but were friends of the chairman of the board of directors. Only the third member seemingly had no direct link with the chairman. Inevitably, the supervisors were reluctant to challenge decisions of the main board. When asked about the power of the supervisors, an interviewee of Silan commented:

> The founders of the company experienced the difficulties and hardships together when they started the business. They are like brothers. Now the company is very successful and the founders have different responsibilities for the company. Two of them sit on the supervisory board. What they could do is to consult or advise at most. How could you imagine them to monitor their brothers mercilessly? (Interview 39)

Many SB members identified themselves as friends and associates of the board of directors, even though the revised Chinese Company Law (2005) stipulated that the SB has the right of overseeing and monitoring the directors, even bringing lawsuits to the court.

A lack of independence

The overwhelming majority of the interview respondents agreed the SB lacked independence. They said the SB was a formality only, entirely subservient to the main board, and unable to monitor effectively the chairman of the main board and the actions of the board in general.

The lack of independence of the SB was at least in part a product of its composition, as noted above. Outsider supervisors were nominally independent and were not employees or from the linked companies. Insider supervisors, conversely, worked for the company or its parent company. They had the advantage of being better informed than outsiders but were likely to lack independence (Xiao et al., 2004). Among the SB members of the case study firms, insider SB members accounted for the majority of members (Table 5.3).

| Table 5.3 | Personnel composition of the supervisory board, 2008 |

	Hisense	Huadong	Panda	Haier	Chunlan	Yankon	Tiantong	Silan
Insider members	3	3	5	3	2	2	3	2
Outsider members	N/A	N/A	N/A	N/A	1	1	2	1

Source: Interviews with board members of the case study firms (2008).

Moreover, most of the outsider SB members of our firms came from the parent company of the listed companies. Outsider specialists in business management were rarely included in the membership of the SB. In this sense, the SB in these firms was less able to act as a supervisory board in the way such boards operate in Germany.

Independent directors

Table 5.4 shows that a recent change in Chinese boards is the introduction of independent directors, which were non-existent when the companies initially listed.

Silan is the exception since it listed after the guideline Establishment of Independent Directors Systems by Listed Companies Guiding Opinion (Guiding Opinion, 2001 hereafter) was published in 2001 in China, and is mentioned in Chapter 3. Independent directors were not required before 2001 (see Table 5.4). In the late 1990s, the trend to appoint independent directors spread from developed market economies to developing countries (Tenev et al., 2002). Acknowledging the importance of a sound corporate governance system, many developing countries made stipulations concerning independent directors of listed firms, including China (Shen and Jia, 2004). In view of the high concentration in ownership of listed firms in China, the introduction of the independent director in firms was designed to prevent controlling shareholders from using their advantageous positions to the detriment of both the corporation and of minority shareholders. However, most independent directors interviewed did not fulfil their monitoring role for several reasons, described below.

| Table 5.4 | | A comparison of boards in case firms at different stages | | | | |

		Size of board of directors	Employee directors	Affiliated directors	Independent directors	Supervisory board size
First year the firms listed	Hisense (1997)	8	5	3	N/A	2
	Huadong (1997)	7	4	3	N/A	3
	Panda (1996)	13	2	11	N/A	3
	Haier (1993)	6	2	4	N/A	3
	Chunlan (1993)	7	5	2	N/A	3
	Yankon (2000)	9	5	4	N/A	3
	Silan (2003)	11	4	2	5	3
	Tiantong (2001)	9	7	2	N/A	5
2007	Hisense	8	4	1	3	3
	Huadong	9	4	2	3	3
	Panda	9	1	6	2	5
	Haier	9	4	3	3	3
	Chunlan	9	5	1	3	3
	Yankon	9	4	3	3	3
	Silan	11	5	1	5	3
	Tiantong	9	5	1	3	3

Notes: Affiliated director: A director who also holds a position in the parent company; Independent director: A director who is neither a shareholder nor an employee and who does not receive shares or salary; Employee director: A director who represents the interests of an enterprise's employees.
First year of list is shown in parentheses after the firm's name.
Source: Annual reports of case study firms (various years).

A lack of information and skill

Analysis of the background of the independent directors of the sample firms shows that most were academic researchers or independent

Table 5.5	Percentage of independent directors from different areas, 2007		

	SOE	COE	DPOE
Academic	44	50	73
Accountant	22	23	9
Experts and professionals	22	21	9
People from unaffiliated firms	12	6	9

Note: 'Experts and professionals' in the Table refers to lawyers, finance professionals or experts who are working in the related trades societies.
Source: Annual reports of the case study firms (2007).

professionals who were invited to sit on the board (Table 5.5). Few seem to have the required expertise to fulfil their role.

The appointment of independent directors to the main board was not only to meet the requirements of the government, but more importantly to provide an alternative source of professional advice to that of the executives of the firm (Interview 7). Independent directors are charged with monitoring management integrity and performance (Guiding Opinion, 2001). However, these academics from universities often lacked time and did not possess specialist knowledge of business. One interviewee commented:

> Selecting many famous academics reflects the preference of the executives of the listed firms. They could meet the required minimum number of independent directors, and also enhance the reputation of the firm. More importantly, these famous people have no time, energy and knowledge to monitor the business. Selecting them as the independent directors will avoid trouble for the controlling shareholder. They are suitable to be the 'vases' of the board room. (Interview 53)

Such decorative appointments as independent directors were hardly likely to serve their intended corporate watchdog function and presented little threat to the interests of large shareholders.

A lack of independence

Article V of the Guiding Opinions (2001) states that the independent directors of a listed company have a legal duty to express their opinions on the company's significant proceedings such as nomination, appointment, and dismissal of directors and senior executives; the compensation of directors and senior executives; and any financial commitment such as large loans (Guiding Opinion, 2001). That is, any major related-party transactions should be approved by the independent directors. However, the independent director system in the case firms apparently has not achieved these expected results. Their independence has been undermined.

Large shareholders who dominated the board nominated the majority of the independent directors appointed (Tong, 2004). In most of our cases, the independent directors who were invited to join the board had a friendly relationship with managers (Interviews 7, 13, 21, 32, 48). Most independent directors were appointed by the very people they were to monitor. The 'Suggestion on Improving Independent Director and Supervisory Board System' (hereafter referred to as Suggestion, 2008) released by CSRC in 2008 admitted that in many listed firms the major shareholders or management selected the independent director. Each year there were various 'under the table' connections with controlling shareholders or management concerning the appointments of independent directors (Suggestion, 2008). Controlling shareholders and management selected those with whom they had connections and who would side with them. One independent director explained his function quite frankly:

> It is extremely rare for the independent director to oppose diametrically the position of the board and management. In most cases the independent directors will turn a blind eye to the decisions made by the board or management. The secret for me to survive and perform my duties in the firm is to veto the simple issue and approve the complicated issue. Since the number of independent directors is fewer than half of the board members, my vote will have absolutely no influence on the final decision. (Interview 32)

In addition, the compensation of independent directors detracted from their independence. Most independent directors earned between RMB 40,000 and RMB 50,000 (about US$5,000–6,000) a year, though a few were volunteers without compensation (Shen and Jia, 2004). An interviewee proudly told us: 'I just hurried back from Australia to attend the board meeting. I had to come back. If I attend all the meetings called by the board, I could earn RMB 60,000 in board fees. That is even higher than my annual salary' (Interview 21). If the compensation was too low, independent directors might lack incentive, but if too high, independence was seemingly eroded. In such situations, the voice of the independent director was weak.

Affiliated directors

In China, non-executive directors are from outside the company and are classified into independent directors and affiliated directors (Tian and Lau, 2001). Since many firms in China listed only some of their business units rather than the entire company, the parent company and its listed firm are affiliated via shared land, production, facilities and staff resources. Compared with independent directors, who are not employed by the parent company and its listed companies, affiliated directors in the listed firms constitute a special social group that has close linkages with the parent holding company, which is the controlling shareholder of the listed entity. Some executives in the holding company serve as board chairs, CEOs or Party Committee secretaries in the listed firm. Although these affiliated directors are not on the payroll of the listed company, many have worked in the same enterprise with most or even all of the top managers of the parent shareholding company (Tian and Lau, 2001). The board of directors in Chinese listed firms is often staffed with individuals that are directly or indirectly affiliated with the controlling shareholder (see Table 5.6).

The prevalence of directors affiliated with the largest shareholder at the initial listing stage was not surprising, given that the majority shareholder typically possesses a large equity stake in the listed firms. Table 5.4 shows that an average listed entity of the case firms had a board of directors with about nine members; of these, about 42 per cent were affiliated directors on average in the first year the firms listed. Panda was the highest in the sample with 78.5 per cent (see Table 5.6).

Table 5.6 Percentage of affiliated directors in the case firms at different stages (%)

	Hisense	Huadong	Panda	Haier	Chunlan	Yankon	Silan	Tiantong
First year of listing	37.5 (1997)	42.8 (1997)	78.5 (1996)	66 (1993)	28.5 (1993)	44.4 (2000)	18.1 (2003)	22.2 (2001)
2007	12.5	22.2	66.6	33	11.1	33.3	9	11.1

Source: Annual Reports (various years).

Affiliated directors were not independent from the management. The close social ties between the affiliated directors and managers of the listed firm implied that the two parties might easily collaborate to pursue personal goals at the expense of the organizational goals (Tian and Lau, 2001). According to Article 14 of the Company Law (2005): 'The company may establish subsidiary companies, which have the status of enterprise legal persons and bear civil liabilities independently in accordance with the law'. In addition, the annual reports of the listed firms asserted that the listed firm was independent of the holding companies in terms of 'personnel, assets and finance' (Annual Reports various years). The overlapping positions, however, implied that the listed firm and holding company were interdependent. A senior consultant observed:

> The listed firm and the holding company are affiliated because the latter continues to provide various services to the former, and the former may rent land or production facilities from the latter. Except for these contract-based business relationships, the two may maintain informal but close relationships over a wide range of areas. (Interview 25)

Table 5.6 shows for 2007 that the average number of the affiliated directors had decreased significantly, except for Panda where affiliated directors were still the majority. With fewer affiliated directors, collusion between representatives of the parent and listed firm's top management at the expense of shareholders was likely to decline (Tian and Lau, 2001).

Conclusion

This chapter explored the steps Chinese firms have taken to improve their corporate governance mechanisms, such as clarifying the structure of ownership and broadening the membership of boards. Figure 5.10 summarizes the main features of the ownership structure and board characteristics found among our eight cases. Examining the case analysis has shed light on our second major research aim: to understand the characteristics of corporate governance of firms in the Chinese CE sector. Below we review our findings concerning changes in ownership structure for each type of firm studied here and conclude with a comparative discussion on understanding how the model of corporate governance is emerging in China.

Figure 5.10 The transformation of ownership structure and board characteristics of the cases at different stages

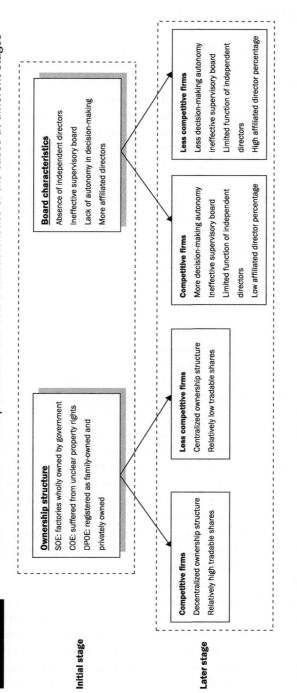

The situation in SOEs

The SOEs increasingly decentralized control. Ownership of the State-controlled listed firms shifted from one in which the State held an absolutely dominant position to one of a multi-party company ownership structure. Private investment was injected into the SOEs through listing part of the firm, though ultimate ownership remained securely in the hands of the State, in either local or central government.

In the examples of the State-controlled listed firms Hisense and Huadong, the State relaxed its ownership control. With the decentralization of ownership from the central government to local government, local governments were able to make a residual claim on the profits of the companies, while they increasingly delegated decision-making power to management. As one moves down the State administrative hierarchy, the goals and interests of local governments are increasingly aligned with those of the local firms. The monitoring and controlling capacities of local governments were strengthened, but they placed more importance on the economic development of the firms. In this situation, SOEs do not subordinate the pursuit of profit to other requirements such as employment and social welfare. Although State-owned, ownership and management were increasingly separated in practice as seen in the autonomy of decision-making without recourse to government approval. One might expect that this trade-off of ownership for more autonomy would result in a transfer of control, or even quasi-privatization. This supposition is not generally true in the case of China. As Fan et al. (2005) point out, since local governments are constrained by Chinese laws that prohibit the free transfer of State ownership, partially listing local State firms allows local governments to decentralize decision rights to firm management without abandoning their ownership. The government was able to prevent the loss of State control when companies went public via this mechanism.

Although the State may take a more open attitude towards the ownership of small and medium-sized SOEs in the CE sector, it is less likely to do so with strategically and politically important firms such as Panda. These SOEs with special status and large scale tend to be associated with a higher level of Party control and a higher level of State shareholding, and a corresponding reluctance to loosen control over these firms. The government is, however, more willing to subsidize these firms. For these firms, the need to guarantee their survival is greater than the need to augment returns on capital. In this situation, although the SOE listed part of the firm, decentralization of ownership has stalled.

The situation in COEs

COEs in China are a large grey zone in ownership form. They are categorized as non-State, despite varying degrees of government influence or control. The two urban COEs studied here, Haier and Chunlan, have roughly the same kind of ownership structure as many State-owned listed firms because during the deepening of the economic reform in China, small and medium-sized COEs were either merged with ailing SOEs or transformed into privately-owned firms. The survivors are leaders in the CE sector. The COEs' importance in the local economy encourages support from local government. The distinction in ownership between large urban COEs and the SOEs is blurred. The absence of State shares shown in a company's annual report, however, does not necessarily indicate that the local government does not exercise control. The legal person share, the dominant form of shares in the COEs, is a veil for various State-connected entities, including local government.

As with State-controlled listed firms, the managers of collectively controlled listed firms gained more autonomy despite the ultimate control of local government. The transformation of the ownership structure of the State and collective firms illustrates the decentralization of control. On the surface, the transformation of SOEs into a modern corporate system was still wanting, with continued State control and the Party and administrative leadership retaining prominent positions on the BoDs. Nevertheless, the gradual transformation of the ownership structure of SOEs and COEs, however 'trivial or mechanical' the initial change, these 'have sowed the seeds for real changes in the future. Once progressive reform gets under way, its smooth progress hinges on the guidance of the market force' (Jiang, 2001c).

The situation in DPOEs

Separation of ownership and management rights are uncommon – even unlikely – in privately owned firms in China. The controlling owners occupied the top managerial and board positions in the listed firms. By introducing different ownership classes through the creation of State shares (held by State firms or agencies) and legal person (LP) shares, the listed companies of DPOEs also evidenced a mixed ownership. Privately controlled listed firms might be controlled via a holding company or directly by individuals. The first type was more common and had many similarities with Chinese family-controlled firms listed outside China. Compared with SOEs and COEs in which the main purpose of partial listing adopted by the local government is to decentralize the firm's managerial decision rights, the primary purpose of listing for the DPOEs is to overcome financial constraints, in particular the unwillingness of the State bank sector to lend to private firms.

Although the corporate governance structures of Chinese firms remain weak, with the main board dominated by insiders, ineffective independent directors and a compliant supervisory board, there have been big strides since the 1990s. There are fewer insiders or affiliated directors than in the past and independent directors have been introduced. With the decentralized ownership structure of firms, in spite of the continued presence of State ownership, boards have become more like those of modern corporations. And the boards in our case studies have acted independently, with perhaps the notable exception of Panda.

Taken together, the evidence in Chapter 5 suggests that firms with very different kinds of ownership – State, collective or private – can be competitive in the market. There is not an optimal ownership and board structure that is common to all firms that wish to experience higher performance. Corporate governance structures vary across firms, and may result in equally good performance. In the Chinese CE sector the role of ownership structure and board structure in improving firm performance is not as direct as some papers expect (Bai et al., 2004; Hu et al., 2004; Qu, 2004; Xu and Wang, 1997). In contrast, the better the performance of the firms, the more advanced the corporate governance mechanism. This finding accords with the view of Kole (1996) that corporate governance mechanisms are endogenous and plausibly determined, among other factors, by firm performance itself.

Comparison of corporate governance models in China

The model of corporate governance that develops in China is likely to embody the special role of State and cultural aspects of China, while taking on some of the characteristics of the Anglo-Saxon and Rhineland models of corporate governance such as a board of directors and a supervisory board (Mallin, 1998). The board of directors in China is styled after the Anglo-Saxon model of corporate governance, where the board oversees and aids management decision-making. Similar to practices followed in the UK and the USA, guidelines issued by the China Securities Regulatory Commission require that at least one-third of the directors on the main board be independent (China Corporate Governance Survey, 2007). The model of corporate governance in China also contains elements of the German–Japanese bank-centred model, in particular the Rhineland variation. The Rhineland model has a mandatory two-tiered structure. China has adopted a quasi two-tier structure of board governance, with both a board of directors and a supervisory board, as is found in Germany. Despite possessing elements with similarities to the corporate governance model in mature market economies, the governance system in China is different from either the Anglo-American or German–Japanese model.

In the UK and the USA, the ownership is widely dispersed, held in the hands of many minority and often small investors (Shleifer and Vishny, 1997). The capital market and independent directors play a vital role in the corporate governance system. As a general rule, corporate boards in the USA and UK are captured by senior management (Jensen, 1993; Mace, 1971). In China, ownership is concentrated in the hands of large shareholders, which are mostly State-owned agencies or enterprises. Also, the ownership structure is highly concentrated; the corporate boards are in the hands of large shareholders or State-owned large shareholders, who dominate the board and turn independent directors into little more than puppets (Tenev et al., 2002). In addition, external mechanisms such as an active market for corporate control that helps to discipline managers and boards in the Anglo-Saxon model were largely absent until quite recently. Therefore, few external levers exercise effective discipline on Chinese companies (Hovey, 2005).

In China, the State plays a strong role in the corporate governance of large firms, almost all of which are State- or partly State-owned. Unlike the two-tier board structure in Germany, Chinese supervisory boards include Communist Party representatives. These agents of the Party-State combine with large State-owned shareholders to make the supervisory board a rubber stamp (Lu et al., 2009). Subordinating many supervisory board members to the direction of the Communist Party nullifies their ability to supervise directors and managers (Tenev et al., 2002). Supervisory boards rarely contest decisions of the main boards or the company executives; their duties are unclear, their power circumscribed and their independence largely lacking. Despite a formal dual board structure, the structure of the corporate boards in China is in effect one-tier (Wei and Geng, 2008).

Notes

1. The central SASAC was created in 2003 as an organization under the State Council to manage capital assets. Its aim is to control the huge amount of State assets dispersed among SOEs. In 2005, the central SASAC controlled 176 large industrial SOEs, most of which satisfied five criteria: they should be essential to national security, natural monopolies, producers of public goods, important natural resources, or 'pillar industries'. Other SOEs are governed by the local branches of SASAC. The central SASAC oversees the local SASAC initiatives to transform and restructure SOE ownership (OECD, 2004).

2. In order to address the banking system's nonperforming loan (NPL) problem, the Chinese government set up four AMCs in 1999: Huarong, Great Wall, Orient AMC and Cinda (Garnaut et al., 2005). One task of these companies was to implement debt-for-equity swaps. They were to buy up the bad debts of the four big State-owned commercial banks and dispose of them over 10 years. The debt-for-equity swap is an exchange of SOE debts owed to State banks for AMC equity shares. As a result, these companies have acquired control over SOE management. The process has alleviated the burden on participating SOEs and, in some cases, has improved their corporate governance (*People's Daily*, 10 August 2005).

Links between institutions, business strategies and corporate governance

It is not the strongest species that survive, nor the most intelligent, but the most responsive to change.

(Darwinian proverb)

Abstract: Based on the analysis and discussion in the previous chapters, this chapter will explore the paths of development of Chinese firms in the transitional environment. It will propose a model to describe the interrelationship between the institutional environment, business strategies, competitive position and corporate governance of the firms in an emerging economy setting. The model helps to reveal three major findings. First, as China's economic reforms have broadened, market transition and reduced government intervention have reshaped the competitive landscape of firms. Second, the decision-making of the managers of some firms has become more sensitive to market forces and to responding to customer demands, while others continue to be heavily dependent on the government and oblivious to market forces. Third, the competitive position of firms has been sharply polarized between those that are market-oriented and competitive and those that are mostly State-dependent and often less competitive.

The model explains the differences of firms in terms of a virtuous circle of positive feedback in response to market stimulus or a vicious circle of defensive and negative feedback that reinforces anti-market orientations associated with their administrative heritage before the advent of economic reforms. We also suggest the trend and factors influencing the development of Chinese firms. The analysis shows that the development of Chinese firms in the context of China's transition economy does not simply replicate the outcomes predicted in the dominant theory, which is that the strategy configuration of a firm is directly influenced by its ownership type.

Key words: institution, strategy, corporate governance, China, firm development.

Introduction

The objective of this chapter is to develop a model regarding the link between various factors including institutional factors, business strategies, competitive position and corporate governance of firms. The analysis of the link allows for identifying and assessing the different development patterns of firms in the Chinese context. In addition, the chapter also summarizes the factors influencing the development of firms and discusses the development trends of firms in China. Theoretical and empirical contributions are also discussed in this chapter. This chapter elaborates on the limitations that may affect the validity or generalizability of the findings.

The remainder of the chapter is organized as follows. The second section deals with a model concerned with firm development. Section three summarizes the factors influencing the firms, which is followed by a review of trends in the development of firms in China's transition context. The fifth and sixth sections look at the contributions and limitations of the study respectively. Concluding remarks are in the final section.

The model of institutional factors, business strategies and corporate governance in China's transition context

This section evaluates the interrelationship between the institutional environment, business strategies, competitive position and corporate governance of firms. The analysis argues that during China's market-oriented transformation the business strategy, competitive position and corporate governance of firms have been interdependent. Incentives for the managers have changed during the transition process. Different incentives lead to different business strategies, which have an impact on firm's competitive position, and which result in different forms of corporate governance. This in turn affects the incentives of managers. Figure 6.1 shows our model: competitive firms have created a virtuous

Figure 6.1 The interrelationship between institutional factors, business strategies and corporate governance

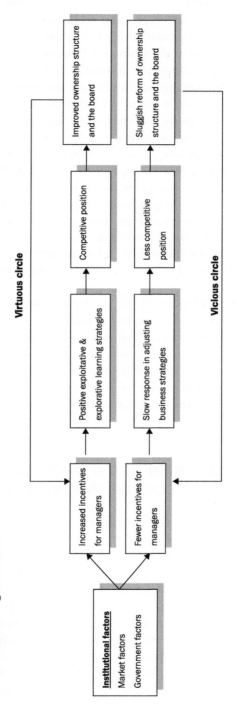

circle for their development path while less competitive firms have become trapped in a vicious circle of attenuated growth.

Market transition in the Chinese CE sector was characterized by the coexistence of two opposing institutional situations: market-oriented elements, and the vestiges of the planned mechanism. On the one hand, the CE sector entered a period of fierce competition and came under more pressure to improve efficiency, which led to a significant and positive effect on firms' adoption of positive learning strategies. On the other hand, bankruptcy as a means for disciplining firms remained rare among China's large SOEs. Some firms, especially large SOEs, continue to exist, despite losing money for years. For this reason, these less competitive firms were dependent on government support and subsidies. Competitive firms, whether State-owned or not, are therefore in a position to differentiate themselves from those that are lacking in their market orientation, learning capability and competitiveness in an environment where the success of a business relies mainly on its strategies and operations.

As a result, polarization has emerged among the firms in terms of competitive position. Competitive firms have grown and continuously expanded their market shares, while the less competitive ones have found survival increasingly tough going. We can explain these differences in terms of a virtuous circle of positive feedback in response to market stimulus or a vicious circle of defensive and negative feedback that reinforced anti-market orientations associated with their administrative heritage before the advent of economic reforms.

The virtuous circle of the development pattern of firms with a competitive position

For competitive firms, the virtuous circle of development is associated with a more market-oriented external environment, which increased the incentives for managers to develop their business. Increased incentives for the managers to pursue profit-seeking goals enabled the managers to become more market- and learning-oriented in their business strategies, which in turn led to firms adopting a more competitive position and resulting in better performance. Since such competitive firms helped the government to achieve their economic goals through employment and taxation revenue, the role of local government in these firms has shifted from the direct administrative control of firms to an indirect role of guidance. This shift has benefitted management's decision-making and responsiveness to the market. It has also allowed for ownership structure to become more diversified and property rights more clarified. This in turn has further strengthened incentives for managers to focus on the competitiveness of firms in the market, thereby initiating another iteration of a virtuous cycle.

The emergence of competitive markets and decentralized government control is two striking features of transformation in the CE sector. Since the emergence of the buyer's market for products where competition is the rule, product market competition has been associated with a relatively high degree of managerial autonomy, which has provided incentives for firms to improve their production efficiency. The findings also suggest that recently, the government has started to value the market-disciplining function implicit in public listing and reduced their interference in the firm that might adversely affect their performance. Central government has gradually streamlined its internal economic sections and shifted decision rights to the local government, which in turn shifted the rights to business enterprises.

What makes local government different in the transformation period is that local government officials depend mainly on the performance of local businesses to demonstrate and valorize their own capabilities. Correspondingly, local government was motivated to allow firms it oversaw to adopt strategies that are more independent, provided there is a credible expectation that the strategies could induce faster growth of firms and higher sales than would be the case if the government intervened in firms. Local government therefore devolved to the management of the firms the rights to devise and implement strategies so as to encourage a response to the market and to maximize the value of the firm. The residual claims were thus transferred from local government to the manager. According to Walder (1995b: 270), 'governments at the lower levels are able to exercise more effective control over their assets than are governments at higher levels'. This decentralization of rights promotes sufficient competition to constrain government intervention in State and non-State firms.

The incentives derived from the environmental dynamism of this institutional transition compel the firms to be learning-oriented in their business development. If the firms were able to respond to the dimension and pace of institutional change, such newly acquired autonomy and flexibility motivated the firms to design strategies to build resources and capabilities to compete. Motivated by profit, the managers responded enthusiastically to opportunities for growth. Since firms have played an increasing role in managerial decision-making during China's transition, managers have exercised their greater freedom to implement the business strategies necessary to satisfy customers. The analysis shows that the firms became learning-oriented in order to survive fierce competition. The changes in business strategy over time reflected their adaptability to the changing structure and opportunities of the market. The analysis shows that in the context of the business environment discussed above, the business strategies of the firm were determined by incentives for the managers. The firm itself was transformed through a series of learning behaviours that sharpened its competitiveness.

This study holds that the better performance of a firm would lead to a dilution of State ownership in SOEs and strengthen the independence of the board of directors. Seemingly, the better their performance the less concentrated is the ownership of firms. This is consistent with the idea of Kole (1996) that corporate performance could influence the ownership structure or board structure rather than being determined by ownership structure. This study points to the endogeneity of corporate governance mechanisms on performance within the Chinese context.

However, this does not simplistically infer that corporate governance determines firm performance in a unidirectional way; we need to make an explicit account of the mutual links between firm performance and the governance mechanism. Superior performance may influence the governance choice of firms. First, in the cases of government-oriented firms, since competitive firms provided substantial benefits to local government they were more successful in improving their bargaining power, providing them with leverage to resist interference from local government. In general, the greater the firm's bargaining power, the greater its ability to reduce government interference. The growing indispensability of the firms has contributed to their influence and created a new power balance in favour of firm managers. When it comes to business decisions, the prestige and reputation of the firms usually carry weight in decision-making, as long as local government officials believe their strategies are sound. Accordingly, government relinquishes its majority stake in the government-oriented firms as maintaining ownership becomes increasingly counterproductive in a competitive marketplace. In terms of DPOEs, this study shows that with improved performance, concentration of ownership in these firms also declined. Some control was transferred from a controlling owner to outsiders when ownership concentration was reduced. Second, listed companies paid more attention to strengthen the functions of their boards, to decrease the number of affiliated directors and to increase the decision-making power of the directors as the ownership structure became progressively less concentrated. Although the board of directors and the supervisory board in these firms were far from operating in an efficient manner, our investigation of the evolution of the board structures of the firms shows that directors were drawn increasingly from among professional practitioners.

A decentralized ownership structure and a more independent board gave managers more control and more rights to claim residual income, which established appropriate incentive mechanisms related to firm performance, thereby starting another iteration of the virtuous circle.

The vicious cycle of the development pattern of firms with a less competitive position

For less competitive firms there were fewer incentives to adopt market-oriented strategies and they became trapped in a vicious cycle of reliance on State support, which reduced their competitiveness over time. These firms expected the past model of business operation still to work in a fast-changing market: governmental authorities or agencies, rather than the market, would primarily determine the fate of the firm. Although the level of government intervention was reduced, some firms were poor in adapting to the market and continued to seek protection from the government. Thus their managers lacked the incentive to adopt positive learning strategies to grow the firm. These firms had weaker aspirations to exploit opportunities in the external environment and to develop production, technological and organizational capabilities, which meant they were less inclined to engage in learning-oriented strategic activities. Sticking to old strategies in a changing environment may lead to poor performance and undermine the firm's competitive position (Audia et al., 2000).

Although our case study SOE Panda was less competitive than the other SOE cases, it was very large and the government wanted to retain a majority shareholding in such a significant State-controlled firm in order to prevent it going out of business. Interference from the government in Panda strengthened as its performance worsened, which further impaired the incentives of the managers. The firm was unable to seize the opportunity provided by the changing institutional and market environment, thereby initiating another iteration of the vicious cycle.

Factors influencing the development of firms in China's transition context

Evidence from this study shows that during the Chinese transition, firms with the same form of ownership structure may behave differently, or firms with different forms of ownership structure may behave similarly. The following presents the common factors influencing the development of firms.

First, the incentives provided by different types of ownership are likely to be similar. Many existing studies hold that privately owned or non-State firms behave differently from government-oriented or State-owned firms because of different incentives. Unlike private firms, government-oriented firms are presumed to pursue maximization of political support rather than profits, and accordingly they have failed to provide incentives for managers to maximize economic performance (e.g. Caves, 1990; Shleifer and Vishny, 1994). Thus managers have lacked the incentive to adopt positive learning strategies to achieve economic benefits. However, this argument is not applicable to all government-oriented firms in China. During the transformation period in China, local government officials have depended on local businesses to demonstrate their capabilities, and correspondingly have increased incentives to monitor local firms' economic performance based on market practice. Facing fierce competition in the CE sector, governments may respond to competitive pressures by lowering their intervention in these firms as it becomes increasingly counterproductive in a competitive marketplace. In this case, local government has preferred to allow the management of the firms to devise and implement decisions so as to encourage a rapid response to changes in the market and maximize the market value of the firm. The residual claims are thus transformed from local government to the manager. The evidence suggests that since measures were brought in to create profit incentives for the firm as a whole, similar to the DPOEs, the government-oriented firms are likely to have the same incentives to use market-oriented learning strategies to grow their business as long as the principal–agent interests are aligned. It can be concluded that if the type of ownership provides distinct incentives for the firm's choice of learning strategies, one would expect significant differences of business strategies based on ownership. But if the effect of the ownership is mediated through some other institutional processes such as competition development, the competitive nature of the industry and decentralization of government control under China's current situation, these differences between firms of different ownership form should be less salient.

Second, in the Chinese transition period, although both business strategy and corporate governance influence the competitive position of the firms, the influence of the business strategy is more direct than that of corporate governance. Since the reform related to corporate governance is relatively harder and more sensitive because it is about who controls and owns the firms, business strategy reform may drive corporate governance reform. More advanced transformation of firms in China always starts with business strategy, which has little or no relationship with the ownership system. However, in time, allowing firms to have more autonomy in decision-making could cause an accumulation of changes in ownership structure. This implies a 'strategic-oriented' approach to the development the firms in transition period.

The reform of one side of the system may penetrate other sides until the entire Chinese economic system completes its transition to the market economy. The similarities of corporate governance between firms with different ownerships are greater than expected. This echoes the idea of Jiang (2001c) that ownership reform is not a magic drug that enables China's CE companies to prosper. Only by seizing the opportunity and improving resource allocation can they survive and develop further in an institutional environment that is more market-oriented.

Third, government shareholding is beneficial only when the firm is competitive. Since the role of State shareholding remains both interfering and supportive, managers from the competitive firms are able to resist the interfering and constraining functions of local government. If companies are less competitive, managers are either heavily dependent on the government for support or unable to eliminate the negative influence of the government. Thus the government as shareholder is detrimental to adjustment to the market when the company is not competitive, as shown in Panda.

The trend of development of Chinese firms

Rawski (2007) argues that it is possible for bureaucratic intervention and market forces to achieve harmony, and that the socio-political and economic rationales could be reconciled. This situation would endure for some time in China with decreasing government intervention and socio-political influence. Firms are allowed to enter the scene relatively freely with good profit-making prospects and with a motive for profit. In the meantime, the prospects and goals of firms can sometimes take precedence over political demands so long as they are beneficial to the local economy. To a great extent, administrative authorities reduce quite considerably the deleterious influence of bureaucratic interference in the firms' decision-making.

Although the transformation of firms in China began by trying to make managers more sensitive to the changing institutional environment and adopt corresponding business strategy that had little or no relations with the ownership system, this does not mean that reform in corporate governance in China is not important. More autonomy in decision-making has formented fundamental changes in ownership structure. Ownership structure, as one mechanism of corporate governance, is the focal point of the discussion. For SOEs and large urban COEs, emerging new firms of these two types are free from historical burdens and have adopted flexible management mechanisms. These firms are more efficient than the old ones that are encumbered with heavy historical burdens and backward management mechanisms. With the growth of the market mechanism, State ownership has become a low-efficiency form of ownership for some enterprises. For DPOEs, although they have shown adaptability and flexibility to the changing environment during the transition process, their governance arrangements are not impeccable. As with their SOE and COE peers, involvement with the government also exists in DPOEs. Improved governance arrangements are necessary to enable them to develop in competitive market conditions (Tam, 2000).

Bearing in mind that reform of corporate governance is not necessarily the only condition for the enterprises' transformation, complementary elements including improved legal frameworks and effective financial control are also required. Tam (2000) stresses that 'neither the establishment of a modern corporate system nor the development of a corporate governance model in China would by itself deliver every reform goal and serve as a panacea for all the problems. The process could serve as a powerful focal point for contemplating and bringing about other related changes in key areas including corporate performance.'

Contributions

This study is exploratory. It seeks to explore business strategies and corporate governance in transitional China. Three aspects of Chinese firms reflect the nature of the Chinese transition economy. Most firms in China, even State-owned firms, now find themselves increasingly involved in the market because the State no longer allocates production inputs nor sells their outputs to prearranged buyers. Meeting the needs of the market has become an important determinant of survival.

This study makes two sets of theoretical and empirical contributions. Theoretically, it integrates the insights from three theoretical perspectives to explain business strategies and deepens our understanding of corporate governance by considering institutional factors in China.

It is important to remember that the shift from a planned and centralized economy to a market economy has never before been undertaken on as big a scale and as fast a pace as in China (Leahy, 2006). Empirically, this study highlights the need to focus on the particular institutional context of the Chinese transition economy.

Theoretical contribution

The contributions of this study, compared to the present state of knowledge, are summarized in Figure 6.2 and are discussed below.

Our study makes several contributions to the theoretical discussion of business strategies. First, the discussion of business strategies in this book moves beyond the typical resource-based view by integrating the insights of institutional theory with the resource-based view and transaction cost economics to study the evolution of learning strategies by which competitive firms in the Chinese CE sector have adapted to the dynamic changes of the institutional environment. With the longitudinal two-phase data, this study focuses on how to exploit firm-specific resources and at the same time develop new ones so as to opitimize fit with the external environment.

Second, the dominant view that government-oriented firms are associated with fewer positive learning strategies should be reconsidered. The advantage that privately-owned firms have over government-oriented firms is not as large as previously believed. Privately owned companies are said to be superior to government-oriented ownership in employing competitive strategies because privately owned firms have a single and clear objective of profit maximization, hard budget constraints and better incentive mechanisms. These determinants may also exist in competitive government-oriented firms. Government-oriented firms are not all alike. Managers of competitive SOEs, which are mostly under control of local government, in general have more managerial autonomy than do managers of firms under the ownership of central government, and act in a more entrepreneurial and market-inspired competitive spirit.

Figure 6.2 Contributions of the study

Compared to previous research, this study has made distinctive theoretical and empirical contributions by examining business strategy and corporate governance in the Chinese consumer electronics sector, integrating the theories related to business strategy and expanding agency theory.

Contribution of the study

Aspects	Present state	New contributions
Business strategies	• Fragmented discussions related to the elements of business strategy • Static analysis of strategies adopted by China's firms	• Integrative view combining IT, RBV, and TCE • A dynamic approach to strategic fit research
Corporate governance	• Stating the unique type of CG in SOEs, COEs and DPOEs • Government-oriented firms suffer more political and agency costs than DPOEs	• The inclusion of institutional theory in discussing corporate governance • Some government-oriented firms also possess a high degree of managerial autonomy
Combination of BS and CG factors	• Mainly focusing on unidirectional links between business strategy and performance or corporate governance and performance	• One of the first attempts to combine research on the most important elements of business strategy, corporate governance and bidirectional links among factors
Policy making	• Favouring privatization and only improving the governance structure	• Positive learning strategies are important for competitive position of all kinds of firms

Third, in discussing business strategies, the study identified the processes by which firms are able to maintain a dynamic strategic fit in a changing environment. The study explicitly incorporates this dynamic perspective into the study of the fit between firms and their changing institutional environments.

Our study also contributes to the theoretical discussion on the nature of enterprise governance. This study argues that firms within the same category of ownership manifest different characteristics of corporate governance. Ownership does not formally ordain governance form. The evidence shown in this study is different from that of the existing literature concerning the categorization of different types of corporate governance of various firms. Most of the existing literature holds that SOEs, COEs and DPOEs reflect unique types of corporate governance. However, Chinese firms during the transition economy display some similar patterns of corporate governance. Both SOEs and large urban COEs can be categorized as conforming to a government-oriented governance pattern. Competitive government-oriented firms, once production units in the centrally planned economy, are now organized with a substantial degree of managerial autonomy and a separation of management from ownership. To a great extent these firms in China are characterized by a separation of management from ownership. In addition, SOEs, COEs and DPOEs share some common key features: different ownership classes, remaining Party/government influence and affiliated directors. One needs to keep in mind that, in China's transition economy, many firms are in the process of transformation from traditional (that is, State) to more complex and less clear-cut ownership structures. The kinds of uncertainty in the conception of ownership regimes are intrinsic to the very transformation processes being studied here.

Finally, our study combines business strategy and corporate governance factors in the discussion of performance, which is underdeveloped in the past literature that has focused on the link between business strategies and performance or corporate governance and performance. The findings suggest that the relationship between corporate governance and performance is not as straightforward as was predicted. The cases used in this study explain why, in China, corporate governance has no apparent effect on firms' performance. Indeed, theoretical arguments on business strategy and corporate governance become incomplete if such an association is not established.

From an economic perspective, this study supports the proposition that business strategy is mainly a function of market and incentive structures rather than of ownership per se. It also lends credence to suggestions that government-oriented firms in many countries are less competitive, not because they are owned by the State, but because of a lack of explicit goals and objectives and because of State demands that can compromise the pursuit of efficiency and profitability (Heracleous, 2001). In this light, private ownership is neither a necessary nor a sufficient condition for market-oriented learning strategies. The cases studied in this research constitute a potent challenge to the widely held view that private ownership is an indispensable prerequisite to market-oriented learning strategies.

Moreover, evidence found in the cases in the context of China's transition economy does not support the dominant theory that the strategy configuration is directly influenced by the ownership type an organization takes (Tan, 1996, 2002). This study argues that government-oriented firms are not necessarily less positive in taking exploitative and explorative learning strategies than the firms with a privately owned type of corporate governance.

Empirical contribution

This study also has some wider policy-related implications related to the reform of Chinese firms. Unlike ownership proponents who favour only privatization, and governance proponents who favour only improving the governance structure, this study believes that both increasing market-based strategies and improving the governance structure are necessary in enhancing the competitive advantage for Chinese firms. There might well be a role for ownership diversity in stimulating competition growth without rapid and large-scale privatization. China has attempted to create market competition without a large-scale privatization. Future reforms should be directed simultaneously towards these two aspects, and any attempt to lean towards one aspect would undoubtedly have unfavourable consequences. This would lead to the selection of a business strategy posture appropriate for that goal, irrespective of the ownership structure of the firm.

Market competition that leads to positive learning strategies is more important to a firm's competitive position than the reform of corporate governance in enhancing the efficiency of firms, regardless of the type of ownership. This is consistent with the view that the most dynamic industrial growth occurs in areas in which the transition to a market economy is closer to completion (Walder, 1995b). The CE sector in China is more dynamic than the controlled industrial sectors (for example, the defence and airline industries) because firms in this sector are exposed more fully to market institution and market competition. These learning strategies may help to overcome the lack of growth in the corporate governance in the sector, which is a legacy of the pre-reform economy.

The CE sector is representative of the Chinese manufacturing industry's transition to the market economy. Quite a few of the front-runners are SOEs in this sector. If the experiences of these firms are disseminated to other SOEs, then Chinese SOEs as a whole can continue to maintain their competitiveness. The empirical analysis of this sector may provide some insight and food for thought regarding the policies of the other industries in China under the economic transition. Since other industrial sectors in China are undergoing a similar transformation from a planned to a market economy, market competition in those sectors will increase with the deepening of the transformation. Therefore, the findings in this book that are applicable to the CE sector also apply to other manufacturing firms competing in China.

Overall, this study adds to research on business strategy and corporate governance, and their relationship in emerging economies undergoing institutional transition.

Limitations

This study is not free of limitations, particularly owing to its exploratory nature. These limitations should be considered when designing any future research project. One limitation is common to any case study, that is, the reliance on a small sample: in this study, eight firms. Given the state of knowledge concerning business strategy and corporate governance in China, case studies nevertheless are an appropriate research methodology through which to understand the change mechanism of the firms. Caution should be exercised, however, when considering the extent to which the results can be generalized. Many issues need to be clarified and explored further in China in industries going through institutional transition.

Another potential limitation is that the firms under study are unable to provide information about the governance structure of the holding parent company. Due to the data constraints, this study focused on the governance structure of the listed firms as reflecting indirectly that of their holding parent companies. There is a critique of this approach. Evidence indicates that Chinese firms always listed their best part (Qu, 2003). These listed firms have survived a screening process and presumably are better managed (Peng, 2004).

Conclusion

This study demonstrates that market competition is playing an increasingly important role in shaping the learning strategies of chinese firms. Competition squeezes the cost-price margin and forces firms to have stronger incentives to develop new products and explore new markets. Therefore, gradually strengthened industrial competition, the introduction of more efficient ownership types and a development in China's industry towards more flexible ownership types may have improved industrial efficiency.

Against this backdrop, the book aimed to contribute to the understanding of business strategy, corporate governance and their relationship among firms with different ownership types in China's transition economy: State-owned, collectively-owned and privately-owned firms[1]. It explored the implications of the integration of the resource-based view, transaction cost economics and institutional theory on learning business strategies and agency theory on understanding corporate governance.

The following chapter concludes the book with suggestions for the future study of business strategy, corporate governance and their links in transition economies, especially in China.

Note

1. Our study has focused on business strategy and corporate governance for firms in a specific sector. We have not explored the civil society dimension of corporate governance. One of the reasons corporate governance more or less works most of the time in the Anglo-American business environment is the legion of non-state actors involve in monitoring firms and their agents, including professional association for directors and company secretaries, shareholders associations and the many lawyers and institutional investors. Some of the systemic problems that affect corporate governance in China stem from the Party-State disallowing or restricting independent organisations. There issues are outside the scope of the book; they are, though, crucial to future reform of corporate governance in China.

7

Conclusions

Abstract: This chapter serves to draw conclusions based on a synthesis of the results reported in preceding chapters. The chapter summarizes the book's findings, draws out the managerial implications, and sheds light on prospects for continued corporate reform in China. We also recognize the limitations of the study. This chapter therefore proposes future research possibilities for studying the fate of loss-making SOEs and considers the generalizability of the findings in view of the sample selection. Lastly, the chapter offers practical recommendations to Western and Chinese managers interested in improving their competitiveness in China's transitional business world.

Key words: strategy, corporate governance, firm, China.

The objective of the final chapter is to summarize significant findings of the study, draw out the managerial implications and make recommendations relating to further research in the subject area.

Summary

The main aim of this book is to explore the business strategy and corporate governance that impact on a firm's performance. These two factors are interrelated. The sample firms the for case studies were selected in the context of a single industrial sector – the consumer electronics sector – in order to minimize the influence of industry and technology on management attitudes and organizational behaviour. Another characteristic of the sample firms is that they included different kinds of ownership since the transition economy in China has given birth to a new diversity of ownership types.

Research and managerial implications

This book contributes to the ongoing body of work in relation to business strategy and corporate governance in China, and some clear results have been found. The findings suggest that in the reform process, the strategies and goals of managers and their complement of resources and capabilities change over time. Managers make strategic choices about what types of activities – exploitation or exploration – they adopt to allocate the resources to create a new fit. The managers should make continuous adaptations until they realize the desired level of competitive position.

Another issue identified in the empirical analysis is that business strategies of China's firms are not greatly determined by their formal corporate governance such as the ownership structure and the composition of the board of directors. The choice of the various corporate governance options appears to be associated with the business strategy employed. Therefore, the business strategy of the firm may be the catalyst that drives corporate governance transformation rather than corporate governance reform being the primary driver of better competitive position.

Several policy recommendations can be drawn from the findings and observations of this book. It is recommended in this study that both encouraging positive learning strategies and improving the governance structure are necessary in enhancing competitive advantage for China's firms. There is a possibility for ownership diversity in stimulating competition and growth without rapid and large-scale privatization. Future reforms should be directed simultaneously towards these two aspects, and any attempt to lean towards one aspect would undoubtedly have unfavourable consequences.

An empirical case-based study of the behaviour of firms in China is still in its infancy. This study contributes to a better understanding of the nature of strategy formulation and corporate governance within China's firms.

Avenues for future research

There are some general limitations in this research, which were discussed in the previous chapter. For these limitations, this section proposes the need for further research related to certain broad issues. The following are aspects that have been identified during the course of this study that can be further analysed in future research.

First, the evidence suggests that a formal ownership type becomes less important in the deployment of business strategies, because the market competition has become fierce enough to eliminate inefficient firms unless they have special ties with the government. Although there are some competitive SOEs, some large inefficient SOEs are still protected by the government due to their special status, resulting from their size, history and products. These SOEs have not closed down, even though they are working at a loss. Future research could address the question: If the Chinese government is trying to maintain the current State-dominant shareholding structure in Chinese public companies, what will be the consequences for the loss-making SOEs? How might continued state support for ailing SOEs affect further enterprise reform?

Second, future research should consider the generalizability of the findings of this study to firms in other transition economies. In fact, in many ways it is difficult to compare the cases of SOE reform in China with those in Eastern Europe and the Soviet Union, not only because of their different approaches but also because of the very different historical, economic and cultural contexts (Buck et al., 2000). To further assess its distinctiveness, the practices of China's firms in these two areas and their antecedents should be compared to samples of companies in other countries.

Third, the sample of firms in this study is limited to large-sized ones in their category of ownership type. However, the nature and outcomes of small and medium-sized firms in China might produce results that differ from the findings in this study. For example, the political effects of local government organizations on firms of different sizes might be significantly different. It will therefore be important to extend the analysis to firms of different sizes in China.

China has a diverse, complex and rapidly changing economy. This study suggests that China's experience with enterprise reform indicates that the reform process differs from what might have been suggested by mainstream management literature. While theory would contend that privatization was necessary for improving and upgrading the competitiveness and market-oriented awareness of firms, this study helps to show that efficiency improvement can be achieved without complete privatization, so long as a firm benefits from effective internal improvement of incentive mechanisms and external market mechanisms.

Bibliography

I Sources of references in Chinese

This section lists both primary and secondary material published in Chinese, including statistical materials, case study references, company materials, and government law and regulations.

Part 1. General information

Asia Pacific. Nanjing Panda Electronic: Plans major overhaul (11 October 2000).

China Economic Times. The capital flow of Haier (13 August 2002).

China Securities newspaper, 27 December 2006.

Chinese Electronics Industry Yearbook (1984) Beijing: Dianzi gongye chubanshe.

Chinese Electronics Industry Yearbook (1986) Beijing: Dianzi gongye chubanshe.

Chinese Electronics Industry Yearbook (2002) Beijing: Dianzi gongye chubanshe.

Chinese Electronics Industry Yearbooks (various years) Beijing: Dianzi gongye chubanshe.

Chinese Household Electronics Report (various years) Snapshots International.

Chinese Intellectual Property Statistics Bureau (2005).

Chinese Statistics Yearbooks (various years) Beijing: Zhongguo tongji chubanshe.

New Finance. Why is Silan successful? (2004) 5, pp. 73–75.

Panda Electronics (2002), from *www.cnii.com.cn*.

People's Daily (10 August 2005).

Remin Daily. The Development of Chunlan (29 November 2000).

Shangyu Daily. Yankon quickens to integrate with the world market (27 June 2005).

Shanghai Securities (2 November 2000).

Accessed at *www.sina.com*. The reform of Tiantong (15 July 2003).

South China Morning Post (13 May 2006).

South China Morning Post. China electronic component firms lead in consumption upgrading (31 August 2005).

Xinhua Financial Network News. China's Nanjing Huadong Electronics to invest in lighting joint ventures (20 November 2003).
Wenhui newspaper. Special issue on Chunlan (28 August 2006).
Zhejiang Daily. The opportunities facing Yankon (19 July 2005).
Hisense Annual Report. Various years.
Huadong Annual Report. Various years.
Panda Annual Report. Various years.
Haier Annual Report. Various years.
Chunlan Annual Report. Various years.
Yankon Annual Report. Various years.
Silan Annual Report. Various years.
Tiantong Annual Report. Various years.

Part 2. Laws and administrative directives reviewed for this book

1) The Company Law of the People's Republic of China (Revised, 2005).
2) The Company Law of the People's Republic of China (Revised, 1999).
3) The Company Law of the People's Republic of China (1994).
4) The Securities Law of the People's Republic of China (Revised, 2005).
5) The Securities Law of the People's Republic of China (1998).
6) Administrative Measures on the Split Share Structure Reform of Listed Companies (2005).
7) Code of Corporate Governance for Listed Companies (2002).
8) Codes of the Supervisory Board in SOEs (2000).
9) Establishment of Independent Director Systems by Listed Companies Guiding Opinion (2001).
10) Standards on Corporate Information Disclosure by Publicly Listed Companies (2002).
11) Tentative Regulations for Supervisory Boards of SOEs (2000).
12) Interim Regulations on the Management of Enterprise State-owned Assets (regulation, 5/2003)
13) Notice Relevant to Pilot Reform of the Segmented Share Structure of Listed Companies (2005).
14) Guiding Opinion on the Share Reform of Listed Enterprises (opinion, 8/2005)
15) Circular on Some Issues Related to the Management of State-held Shares in the 15) Reform of Listed Enterprises" Non-tradable Shares (*circular*, 9/2005)
16) Suggestion on Improving Independent Director and Supervisory Board System (8/2008)

II Sources of references in English

This section contains the consulted theoretical works in the management literature as well as secondary materials on Chinese economic development, corporate governance and other aspects.

Agrawal, A. and Knoeber, C. (1996) Firm performance and mechanisms to control agency problems between managers and shareholders. *Journal of Financial and Quantitative Analysis,* 31, pp. 377–97.

Aguilera, R.V. and Jackson, G. (2003) The cross-national diversity of corporate governance: dimensions and determinants. *Academy of Management Review,* 28(3), pp. 447–65.

Aharoni, Y. (1981) Managerial discretion. In R. Vernon and Y. Aharoni (Eds.), *State-Owned Enterprise in the Western Economies* (pp. 184–193). New York: St. Martin's Press.

Allen, F., Qian, J. and Qian, M. (2002) Law, finance, and economic growth in China. Working Paper, Finance Department, University of Pennsylvania.

Ambler, T. and Wang, X. (2002) Measures of marketing success: A comparison between China and the United Kingdom. *Asia Pacific Journal of Management,* 20, pp. 267–81.

Antonelli, C. (2004) To use or to sell technological knowledge. Working Paper, Department of Economics, Di Torino University.

Asia Port Daily News. Hisense expanding marketing channels in Beijing (4 January 2000).

Audia, P., Locke, E. and Smith, K. (2000) Paradox of success: An archival and a laboratory study of strategic persistence following radical environmental change. *Academy of Management Journal,* 43, pp. 837–53.

Backman, M. (1999) *Asian Eclipse: Exposing the Dark Side of Business in Asia.* Singapore: John Wiley & Sons (Asia) Pte Ltd.

Bai, C., Liu, Q., Lu, J., Song, F. and Zhang, J. (2004) Corporate governance and market valuation in China. *Journal of Comparative Economics,* 32, pp. 599–616.

Barney, J. (1991) Firm resources and sustained competitive advantage. *Journal of Management,* 17, pp. 99–120.

Barney, J. (2001) Is the resource-based 'view' a useful perspective for strategic management research? Yes. *Academy of Management Review,* 26(1), pp. 41–56.

Barnhart, S.W., Marr, M.W. and Rosenstein, S. (1994) Firm performance and board composition: Some new evidence. *Managerial and Decision Economics,* 15(4, Special Issue), pp. 329–40.

Baron, D. (1995) Integrated strategy: market and nonmarket components. *California Management Review,* 37(2), pp. 47–65.

Baron, D. (1997) Integrated strategy, trade policy and global competition. *California Management Review,* 39(2), pp. 145–69.

Baysinger, B. and Hoskisson, R.E. (1990) The composition of boards of directors and strategic control: effects on corporate strategy. *Academy of Management Review,* 15(1), pp. 72–87.

Bebchuk, L.A. and Roe, M.J. (2004) A theory of path dependence in corporate ownership and governance. In J.N. Gordon and M.J. Roe (Eds.), *Convergence and Persistence in Corporate Governance* (pp. 69–113). Cambridge, UK: Cambridge University Press.

Becht, M. and Roël, A. (1999) Blockholdings in Europe: An international comparison. *European Economic Review,* 43(4–6), pp. 1049–56.

Beekun, R.I., Stedham, Y. and Young, G.J. (1998) Board characteristics, managerial controls and corporate strategy: A study of US hospitals. *Journal of Management,* 24(1), pp. 3–19.

Berle, A.A. and Means, G.C. (1932) *The Modern Corporation and Private Property*. New York: Macmillan Company.

Bian, Y. and Zhang, Z. (2006) Explaining China's emerging private economy: Sociological perspectives. In A.S. Tsui, Y. Bian and L. Cheng (Eds.), *China's Domestic Private Firms* (pp. 25–39). Armonk, NY: M.E. Sharpe.

Blair, M. (1995) *Ownership and Control: Rethinking Corporate Governance for the Twenty-first Century*. Washington, D.C.: Brookings Institute.

Boardman, A.E. and Vining, A.R. (1989) Ownership and performance in competitive environments: A comparison of the performance of private, mixed and state-owned enterprises. *Journal of Law and Economics*, 32, pp. 1–33.

Boisot, M. and Child, J. (1988) The iron law of fiefs: bureaucratic failure and the problem of governance in the Chinese Economic reforms. *Administrative Science Quarterly*, 33, pp. 507–27.

Boisot, M. and Child, J. (1996) From fiefs to clans and network capitalism: Explaining China's emerging economic order. *Administrative Science Quarterly*, 41(4), pp. 600–28.

Bower, J. (1970) *Managing the Resources Allocation Process*. Boston, MA: Harvard University Press.

Brandt, L. and Zhu, S.C. (2005) Technology adoption and absorption: The case of Shanghai firms. Working Paper, Department of Economics, University of Toronto.

Bruton, G.D., Lan, H. and Lu, Y. (2000) China's Township and Village Enterprise: Kelon's competitive edge. *The Academy of Management Executive*, 14(1), pp. 19–29.

Buck, T., Filatotchev, I., Nolan, P. and Wright, M. (2000) Different paths to economic reform in Russia and China: Causes and consequences. *Journal of World Business*, 35(4), pp. 379–400.

Caldart, A. and Ricart, J. (2007) Corporate strategy: An agent-based approach. *European Management Review*, 4, pp. 107–20.

Campbell-Hunt, C. (2000) What have we learned about generic competitive strategy? A meta-analysis. *Strategic Management Journal*, 21(2), pp. 127–54.

Carrasco, V. (2005) Corporate board structure, managerial self-dealing, and common agency. Working Paper: Department of Economics, PUC-Rio.

Caves, R. (1990) Lessons from privatization in Britain: State enterprise behaviour, public choice, and corporate governance. *Journal of Economic Behaviour and Organization*, 13, pp. 145–69.

CCIDConsulting Report (2007) 2006–2007 Annual Report on the development of China's consumer electronics industry.

CCIDConsulting Report (2008) *http://www.wjcf.net/report_sample/english/673.pdf*.

Chai, J.C.H. (1997) *Transition to a Market Economy*. Oxford: Clarendon Press.

Chang, E. and Wong, S. (2003) Managerial discretion and firm performance in China's listed firms. Working Paper: Faculty of Business and Economics, University of Hong Kong.

Chang, H. and Singh, A. (1997) Can large firms be run efficiently without being bureaucratic? *Journal of International Development*, 9(6), pp. 865–75.

Charkham, J. (1994) *Keeping Good Company: A Study of Corporate Governance in Five Countries.* Oxford: Clarendon Press.

Charkham, J. (1995) *Keeping Good Company: A Study of Corporate Governance in Five Countries.* New York: Clarendon Press.

Charles, O. (1994) One board or two? *Corporate Finance,* 113(April), pp. 40–5.

Chen, J. (2005a) *Corporate Governance in China.* Oxon: Routledge.

Chen, J. (2005b) Corporatisation of China's state-owned enterprises and corporate governance. In D. Brown and A. MacBean (Eds.), *Challenges for China's Development* (pp. 58–71) London; New York: Routledge.

Chen, L.F. (1999) The case study of Township and Village Enterprises. *Kexue Guanli Yanjiu (Scientific Management Research),* 17(2), pp. 67–9.

Chen, M.J. (2001) *Inside Chinese Business: A Guide for Managers Worldwide.* Boston, MA: Harvard Business School Press.

Chen, W. (2007) Does the color of the cat matter? The red hat strategy in China's private enterprises. *Management and Organization Review,* 3(1), pp. 55–80.

Cheng, J.W. and Ma, Z.Y. (7 December 2006) The development of family-owned firms in China. *Zhongguo Zhengquan Bao. China Securities* newspaper.

Cheng, W. and Lawton, P. (2005) SOEs reform from a governance perspective and its relationship with the privately owned publicly listed corporation in China. In D. Brown and A. MacBean (Eds.), *Challenge for China's Development: An Enterprise Perspective* (pp. 24–47) London; New York: Routledge.

Cheung, S. and Chan, B.Y. (2004) Corporate governance in Asia. *Asia Pacific Development Journal,* 11(2), pp. 1–31.

Child, J. (1990) The character of Chinese enterprise management. In J. Child and M. Lockett (Eds.), *Advances In Chinese Industrial Studies,* pp. 137–52. London: JAI Press InC.

Child, J. (1994) *Management in China during the Age of Reform.* Cambridge: Cambridge University Press.

Child, J. and Lu, Y. (1990) Industrial decision-making under China's reform, 1985–1988. *Organizational Studies,* 11, pp. 321–51.

Child, J. and Lu, Y. (1996) Institutional constraints on economic reform: The case of investment decisions in China. *Organization Science,* 7(1), pp. 60–77.

Child, J. and Pleister, H. (2003) Governance and management in China's private sector. *Management International,* 7(3), pp. 13–23.

China Corporate Governance Survey (2007) HK: Centre for Financial Market Integrity.

China Internet Information Centre (2004) Review of the marketization reform of Chinese enterprises. *http://www.fdi.gov.cn.*

China.org.cn (7 November 2003) Market-oriented reforms of China's enterprises in retrospect. *www.china.org.cn.*

Chow, G.C. (2002) *China's Economic Transformation.* Malden, MA: Blackwell Publishers.

Clarke, D. (2003) Corporate governance in China. *China Economic Review,* 14(4), pp. 494–507.

Coase, R. (1937) The nature of the firm. *Economica,* 4, pp. 386–405.

Conner, A. (1991) To get rich is precarious: Regulation of private enterprise in the People's Republic of China. *Journal of Chinese Law,* 5(1), pp. 9–15.

Cook, K. (1995) *AMA Complete Guide to Strategic Planning for Small Business.* Chicago: American Marketing Association.

Crawford, R. and Feng, L. (2000) ECCH: European Case Clearing House. Case Number: 300-129-1.

Crossan, M., Lane, H. and White, R. (1999) An organizational learning framework: From institution to institution. *Academy of Management View,* 24, pp. 522–37.

Dahya, J., Karbhari, Y. and Xiao, J.Z. (2003) The supervisory board in Chinese listed companies: Problems, causes, consequences and remedies. In M. Warner (Ed.), *The Future of Chinese Management*, pp. 118–137. London: Frank Cass.

Dalton, D., Daily, C., Johnson, J.L. and Ellstrand, A.E. (1999) Number of directors and financial performance: A meta-analysis. *Academy of Management Journal*, 42, pp. 674–86.

Davidson, L.H. (1994) On the government of companies. *Corporate Governance: An International Review,* 2(1), pp. 5–7.

Demsetz, H., (1983) The structure of ownership and the theory of the firm, *Journal of Law & Economics*, 26(2), pp. 375–90.

Demsetz, H. and Lehn, K. (1985) The structure of corporate ownership: Causes and consequences. *Journal of Political Economy,* 93, pp. 1155–77.

Demsetz, H. and Villalonga, B. (2001) Ownership structure and corporate performance. *Journal of Corporate Finance,* 7, pp. 209–33.

Denis, D.K. (2001) Twenty-five years of corporate governance research . . . and counting. *Review of Financial Economics,* 10(3), pp. 191–212.

Denis, D. and Kruse, T. (2000) Managerial discipline and corporate restructuring following performance declines. *Journal of Financial Economics,* 55, pp. 391–424.

Denis, D.K. and McConnell, J.J. (2003) International corporate governance. *Journal of Financial and Quantitative Analysis,* 38(1), pp. 1–36.

Desarbo, W., Benedetto, C., Song, M. and Sinha, I. (2005) Revisiting the miles and snow strategic framework: Uncovering interrelationships between strategic types, capabilities, environmental uncertainty, and firm performance. *Strategic Management Journal,* 26, pp. 47–74.

DFAT (2002) *Changing Corporate Asia.* Canberra: Department of Foreign Affairs and Trade.

Dickson, B. (2003) *Red Capitalists in China: The Party, Private Entrepreneurs, and Prospects for Political Change.* Cambridge; New York: Cambridge University Press.

Dickson, B. (2007) Integrating wealth and power in China: The Communist Party's embrace of the private sector. *China Quarterly,* 192, pp. 827–54.

Dierickx, I. and Cool, K. (1989) Asset stock accumulation and sustainability of competitive advantage. *Management Science,* 35, pp. 1504–10.

Ding, X. (1998) Reform of China's state-owned enterprises and its system of property rights. In R. Duncan and Y. Huang (Eds.), *Reform of State-Owned Enterprises in China: Autonomy, Incentive and Competition,* pp. 37–50. Canberra: NCDS Asia Pacific Press.

Dougherty, S., Herd, R. and He, P. (2007) Has a private sector emerged in China's industry? Evidence from a quarter of a million Chinese firms. *China Economic Review,* 18, pp. 309–34.

Duncan, R. and Huang, Y. (Eds.). (1998) *Reform of State-Owned Enterprises in China: Autonomy, Incentive and Competition*. Canberra: NCDS Asia Pacific Press.

Duysters, G., Jacob, J., Lemmens, C. and Hu, J.T. (2008) Internationalization and technological catching up of emerging multinationals: A case study of China's Haier Group. Working Paper: Eindhoven Centre for Innovation Studies, Netherlands.

Eccles, R. (1991) The performance measurement manifesto. *Harvard Business Review, January–February*, pp. 131–37.

Economist, The. Business: Try cabbages next time; Chinese television makers. (2001), 361(8244), p. 93.

Eisenhardt, K.M. (1989) Building theories from case study research. *Academy of Management Review*, 14(4), pp. 532–50.

Eisenhardt, K.M. and Martin, J.A. (2000) Dynamic capabilities: what are they? *Strategic Management Journal*, 21, pp. 1105–21.

Estrin, S. (2002) Competition and corporate governance in transition. *Journal of Economic Perspectives*, 16(1), pp. 101–24.

Fama, E.F. (1980) Agency problems and the theory of the firm. *Journal of Political Economy* 88(2), pp. 288–307.

Fama, E.F. and Jensen, M.C. (1983) Separation of ownership and control. *Journal of Law and Economics, XXVI*, pp. 301–25.

Fan, G. (2000) The dynamic process of institutional transition in China. Jingji Yanjiu. *Economic Research Journal*, 1, pp. 11–24.

Fan, J.P.H., Wong, T.J. and Zhang, T. (2005) The emergence of corporate pyramids in China. Working Paper: Chinese University of Hong Kong.

Feinerman, J. (2007) New hope for corporate governance in China? *China Quarterly*, 191(September), pp. 590–612.

Filatotchev, I. and Toms, S. (2003) Corporate governance, strategy and survival in a declining industry: A study of UK cotton textile companies. *Journal of Management Studies*, 40(4), pp. 875–920.

Financial Times, 15 December 2004.

First Financial Daily, 1 December 2004.

Foss, K. and Foss, N. (2006) Entrepreneurship, transaction costs, and resource attributes. *International Journal of Strategic Change Management*, 1(1–2), pp. 53–60.

Foss, N. (2002) The strategy and transaction costs nexus: Past debates, central questions, and future research possibilities. Working Paper: Department of Industrial Economics and Strategy, Copenhagen Business School.

Foss, N. and Mahnke, V. (2002) Advancing research on competence, governance and entrepreneurship. In N. Foss and V. Mahnke (Eds.), *Competence, Governance, and Entrepreneurship: Advances in Economic Strategy Research*, pp. 1–18. Oxford: Oxford University Press.

Franks, J. (2000) Corporate governance in Germany: The role of banks and ownership concentration: Discussion. *Economic Policy: A European Forum*, 31, pp. 263–5.

Franks, J. and Mayer, C. (1992) *Corporate Control: A Synthesis of International Evidence*. London: London Business School.

Franks, J. and Mayer, C. (2001) Ownership and control of German corporations. *Review of Financial Studies,* 14(4), pp. 943–77.

Fukao, M. (1995) *Financial Integration, Corporate Governance, and the Performance of Multinational Companies.* Washington, D.C.: Brookings Institution Press.

Galbraith, J.R. (1973) *Designing Complex Organizations.* Reading: MA: Addison-Wesley.

Gao, Y.Q. and Tian, Z.L. (2006) How firms influence the government policy decision-making in China. *Singapore Management Review,* 28(1), pp. 73–86

Garnaut, R., Song, L., Tenev, S. and Yao, Y. (2005) *China's Ownership Transformation.* Washington, DC: World Bank.

Garnaut, R., Song, L., Yao, Y. and Wang, X. (2001) *Private Enterprise in China.* Canberra: Asia Pacific Press.

Gatignon, H. and Xuereb, J.M. (1997) Strategic orientation of firm and new product performance. *Journal of Marketing Research,* 34, pp. 77–90.

Ghauri, P. and Gronhaug, K. (2005) *Research Methods in Business Studies: A Practical Guide.* New York: Financial Times Prentice Hall.

Goergen, M. (1998) *Corporate Governance and Financial Performance: A Study of German and UK Initial Public Offerings.* Cheltenham UK: Edward Elgar Publishing Limited.

Gordon, J.N. and Roe, M.J. (Eds.). (2004) *Convergence and Persistence in Corporate Governance.* Cambridge: Cambridge University Press.

Gorton, G. and Schmid, F. (2000) Class struggle inside the firm: A study of German co-determination. Working Paper: National Bureau of Economic Research, Cambridge, MA.

Gospel, H. and Pendleton, A. (2003) Finance, corporate governance and the management of labor: A conceptual and comparative analysis. *British Journal of Industrial Relations,* 41(3), pp. 557–82.

Grant, R.M. (1991) The resource-based theory of competitive advantage: Implications for strategy formulation. *California Management Review,* 33(3), pp. 114–35.

Grant, R.M. (1999) *Contemporary Strategy Analysis.* Malden MA: Blackwell Publishers Ltd.

Green, S.P. (2003) *China's Stock Market: A Guide to Its Progress, Players and Prospects.* London: *The Economist* in Association with Profile Books.

Gregory, N., Tenev, S. and Wagle, D. (2000) *China's Emerging Private Enterprises.* Washington, DC: International Finance Corporation.

Groves, T., Hong, Y., McMillan, J. and Naughton, B. (1994) Autonomy and incentives in China's state enterprises. *Quarterly Journal of Economics,* 109, pp. 183–209.

Guan, Z., Shao, J. and Shen, D. (2001) Tiantong: the first 'natural person' in the Chinese stock market. *Zhejiang Daily.*

Gummesson, E. (2000) *Qualitative Methods in Management Research* (Vol. 2). London: Sage Publications, Inc.

Guo, S. (2007) Designing market socialism: trustees of state property. *Journal of Policy Reform,* 8(3), pp. 207–24.

Gupta, A.K., Smith, K.G. and Shalley, C.E. (2006) The interplay between exploration and exploitation. *Academy of Management Journal,* 49, pp. 693–706.

Haggard, S. and Huang, Y.S. (2008) The political economy of private-sector development in China. In L. Brandt and T.G. Rawski (Eds.), *China's Great Economic Transformation,* pp. 337–74. New York: Cambridge University Press.

Hart, O., Shleifer, A. and Vishny, R.W. (1997) The proper scope of government: Theory and an application to prisons. *Quarterly Journal of Economics,* 112, pp. 1127–61.

Hassard, J., Sheehan, J., Zhou, M., Terpstra-Tong, J. and Morris, J. (2007) *China's State Enterprise Reform.* London and New York: Routledge.

Hawley, J.P. and Williams A.T. (1996) Corporate governance in the United States: The rise of fiduciary capitalism. Working Paper: Saint Mary's College of California, School of Economics and Business Administration.

Henisz, W.J. and Delios, A. (2002) Learning about the institutional environment. In P. Ingram and B.S. Silverman (Eds.), *The New Institutionalism in Strategic Management,* pp. 339–372. Oxford: Elsevier Science Ltd.

Heracleous, L. (2001) State ownership, privatization and performance in Singapore: An exploratory study from a strategic management perspective. *Asia Pacific Journal of Management,* 18, pp. 69–81.

Hitt, M.A., Lee, H.-u. and Yucel, E. (2002) The importance of social capital to the management of multinational enterprises: relational networks among Asian and Western firms. *Asia Pacific Journal of Management,* 19, pp. 353–72.

Holmqvist, M. (2004) Experiential learning processes of exploitation and exploration within and between organizations: An empirical study of product development. *Organization Science,* 15(1), pp. 70–81.

Holz, C. and Lin, Y.-M. (2001) The 1997–1998 break in industrial statistics: Facts and appraisal. *China Economic Review,* 12, pp. 303–16.

Hoskisson, R., Eden, L., Lau, C.M. and Wright, M. (2000) Strategy in emerging economies. *Academy of Management Journal,* 43(3), pp. 249–67.

Hovey, M. (2004) Corporate governance in China: An empirical study of listed firms. PhD thesis, Griffith Business School.

Hovey, M. (2005) Corporate governance in China: An analysis of ownership changes after the 1997 announcement. Working Paper: University of Southern Queensland.

Hu, G. and Georgen, M. (2001) A study of ownership concentration, control and evolution of Chinese IPO companies. Accessed at SSRN: *http://ssrn.com/abstract=286612.*

Hu, Y., Song, F. and Zhang, J. (2004) Competition, ownership, corporate governance and enterprise performance: Evidence from China. Working Paper: University of Hong Kong.

Huberman, M. and Miles, M.B. (Eds.). (2002) *The Qualitative Researcher's Companion.* London: Sage Publications.

Huchet, J.F. (1997) The China circle and technological development in the Chinese electronics industry. In B. Naughton (Ed.), *The China Circle: Economics and Electronics in the PRC, Taiwan, and Hong Kong,* pp. 254–85. Washington, DC: Brookings Institution Press.

Huchet, J.F. and Richet, X. (2002) Between bureaucracy and market: Chinese industrial groups in search of new forms of corporate governance. *Post-Communist Economics,* 14(2), pp. 160–201.

International Finance Corporation (2000) *China's Emerging Private Enterprises*, Washington, DC: International Finance Corporation.

Ireland, R.D., Hitt, M.A. and Vaidyanath, D. (2002) Alliances management as a source of competitive advantage. *Journal of Management*, 28(3), pp. 413–46.

Isobe, T., Makino, S. and Montgomery, D. (2004) Exploitation, exploration, and firm performance: The case of small manufacturing firms in Japan. Working Paper: University of Marketing and Distribution Sciences, Japan.

Jefferson, G., Hu, A., Guan, X. and Yu, X. (2003) Ownership, performance, and innovation in China's large and medium-size industrial enterprise sector. *China Economic Review*, 14, pp. 89–113.

Jefferson, G. and Rawski, T.G. (1994) Enterprise reform in Chinese industry. *Journal of Economic Perspective*, 8(2), pp. 47–70.

Jefferson, G.H., Rawski, T.G. and Zheng, Y. (1996) Chinese industrial productivity: Trends, measurement issues, and recent developments. *Journal of Comparative Economics*, 23(2), pp. 146–80.

Jefferson, G.H. and Singh, I. (1999) *Enterprise Reform in China: Ownership, Transition and Performance*, Oxford: Oxford University Press.

Jefferson, G. and Su, J. (2006) Privatisation and restructuring in China: Evidence from shareholding ownership, 1995–2001. *Journal of Comparative Economics*, 34, pp. 146–66.

Jenkinson, T. and Mayer, C. (1992) The assessment: Corporate governance and corporate control. *Oxford Review of Economic Policy*, 8(3), pp. 1–10.

Jensen, M.C. (1986) Agency costs of free cash flow, corporate finance and takeovers. *American Economic Review*, 76, pp. 323–9.

Jensen, M.C. and Meckling, W.H. (1976) Theory of the firm: Managerial behavior, agency costs and ownership structure. *Journal of Financial Economics*, pp. 305–60.

Jensen, M.C. (1993) The modern Industrial Revolution, exit, and the failure of internal control systems. *Journal of Finance*, 48, pp. 831–80.

Jiang, X. (2001a) Economic transition and industrial development: Relevance, rationale, and significance to theories on economic transition. In X. Jiang (Ed.), *China's Industries In Transition* (pp. 1–56). Huntington, NY: Nova Science Publishers, Inc.

Jiang, X. (2001b) Industrial development and industrial policy in economic transition: A case study of household electrical appliance industry. In X. Jiang (Ed.), *China's Industries In Transition* (pp. 167–197). Huntington, NY: Nova Science Publishers, Inc.

Jiang, X. (Ed.). (2001c) *China's Industries in Transition: Organizational Change, Efficiency Gains, and Growth Dynamics*. Huntington, NY: Nova Science Publishers, Inc.

Jiang, Z. (1997) Jiang Zemin's report to the 15th CCP Congress. *Beijing Review*, October (6–12), pp. 10–33.

Johnson, J.L., Daily, C.M. and Ellstrand, A.E. (1996) Boards of directors: A review and research agenda. *Journal of Management*, 22(3), pp. 409–38.

Johnson, H. and Kaplan, R. (1987), *Relevance Lost: The Rise and Fall of Management Accounting*. Boston, MA: Harvard Business School Press.

Joseph, J. (2006) Haier, the Chinese global competitor. *European Case Clearing House, Reference No. 306-078-1*.

Kang, Y. and Cheng, M.X. (23 February 2009) The independence of boards for Chinese state-owned firms. Accessed at *Economic Observer Online* (*www.eeo.com.cn*).

Kaplan, R.S. and Norton, D.P. (1992) The balanced scorecard-measures that drive performance. *Harvard Business Review, January–February*, pp. 71–9.

Karim, S. and Mitchell, W. (2000) Path-dependent and path-breaking changes: Configuring business resources following acquisitions in the U.S. medical sector, 1978–1995. *Strategic Management Journal*, 21, pp. 1061–81.

Katz, R. (1997) *The Human Side of Managing Technological Innovation*. New York: Oxford University Press.

Keister, L.A. (2000) *Chinese Business Groups, the Structure and Impact of Interfirm Relations during Economic Development*. New York: Oxford University Press.

Khan, A. (2005) Haier in 2005. European Case Clearing House. Case Number: 305-465-1.

Kiran, V.B. and Chaudhuri, S.K. (2004) Haier, developing a global brand. ECCH: European Case Clearing House. Case Number: 304-264-1.

Kohli, A.K. and Jaworski, B.J. (1990) Market orientation: The construct, research propositions, and managerial implications. *Journal of Marketing*, 54, pp. 1–18.

Kole, S. (1996) Managerial ownership and firm performance: Incentives and rewards? *Advances in Financial Economics*, 2, pp. 119–49.

Kornai, J. (1979) Resource-constrained versus demand-constrained systems. *Econometrica*, 47, pp. 801–19.

Kornai, J. (1980) *Economics of Shortage*. Amsterdam: North-Holland.

Kornai, J. (1990) *The Road to a Free Economy. Shifting from a Socialist System: The Example of Hungary*. New York: W.W. Norton.

Kose, J. and Senbet, L.W. (1998) Corporate governance and board effectiveness. *Journal of Banking and Finance*, 22, pp. 371–403.

Kyriakopoulos, K. and Moorman, C. (2004) Tradeoffs in marketing exploitation and exploration strategies: The overlooked role of market orientation. *International Journal of Research in Marketing*, 21, pp. 219–40.

La Porta, R., Lopez-de-Silanes, F., Shleifer, A. and Vishny, R. (1998) Law and finance. *Journal of Political Economy*, 106, pp. 1113–55.

Lardy, N.R. (1996) The role of foreign trade and investment in China's economic transformation. In A.G. Walder (Ed.), *China's Transitional Economy* (pp. 103–120). Oxford: Oxford University Press.

Lardy, N.R. (2002) *Integrating China into the Global Economy*. Washington, DC: Brookings Institution Press.

Lawrence, P.R. and Lorsch, J.W. (1967) *Organization and Environment: Managing Differentiation and Integration*: Homewood, IL: Irwin.

Leahy, C. (2006) Starting down the long road. *China–Britain Business Review, February*, pp. 16–17.

Leask, G. (2004) Strategic groups and the resources based view: Natural complements enhancing our understanding of the competitive process. Working Paper: Aston Business School, Aston University, Birmingham, UK.

Leech, D. and Leahy, J. (1991) Ownership structure, control type classifications and the performance of large British companies. *Economic Journal*, 101, pp. 1418–37.

Levinthal, D.A. and March, J.G. (1993) The myopia of learning. *Strategic Management Journal*, 14(Winter), pp. 95–112.

Li, J. (1994) Ownership structure and board composition: A multi-country test of agency theory predictions. *Managerial and Decision Economics*, 15(4), pp. 359–68.

Li, K. (2001) The independent director system must be improved. *Securities Market Weekly*, 15 October.

Li, M. and Wong, Y.-y. (2003) Diversification and economic performance: An empirical assessment of Chinese firms. *Asia Pacific Journal of Management*, 20, pp. 243–65.

Li, M. and Zhou, H. (2005) Knowing the business environment: The use of non-market-based strategies in Chinese local firms. *Ivey Business Journal*, *November/December*, pp. 1–5.

Li, S. (2004) The puzzle of firm performance in China: An institutional expansion. *Economics of Planning*, 37, pp. 47–68.

Li, S., Li, S. and Zhang, W. (2000) The road to capitalism: Competition and institutional change in China. *Journal of Comparative Economics*, 28, pp. 269–92.

Li, X. (2004) Panda mobile is looking for a 'partner'. *The Economy of Knowledge*, 6, pp. 35–7.

Li, Y.M. (2001) The success experience of Huadong. *Chinese Brandnames*, 4, p. 39.

Lieberson, S. (1992) Small Ns and big conclusions: An examination of the reasoning in comparative studies based on a small number of cases. In C.C. Ragin and H.S. Becker (Eds.), *What Is A Case? Exploring The Foundations of Social Inquiry* (pp. 105–118). Cambridge: Cambridge University Press.

Lin, C. (2001) Corporatisation and corporate governance in China's economic transition. *Economics of Planning*, 34, pp. 5–35.

Lin, J.Y., Cai, F. and Li, Z. (1994) *The China Miracle: Development Strategy and Economic Reform*. Shanghai: Sanlian Press.

Lin, Y.M. and Zhu, T. (2001) Ownership restructuring in Chinese state industry: An analysis of evidence on initial organizational changes. *China Quarterly*, 166, pp. 305–41.

Linz, S.J. (1997) Russian firms in transition: Champions, challengers and chaff. *Comparative Economic Studies XXXIX*, 2, Summer, pp. 1–36.

Listed Company Association (21 March 2003, 2 July 2007) Accessed at *http://www.zjlca.com*.

Liu, G. and Sun, P. (2003) Identifying ultimate controlling shareholders in Chinese public corporations: An empirical survey. Working Paper: Royal Institution of International Affairs, UK.

Liu, G.S. and Garino, G. (2001a) China's two decades of economic reform. *Economics of Planning*, 34, pp. 1–4.

Liu, G.S. and Garino, G. (2001b) Privatization or competition. *Economics of Planning*, 34, pp. 37–51.

Liu, G.S. and Sun, P. (2005) China's public firms: How much privatization. In S. Green and G.S. Liu (Eds.), *Exit the Dragon? Privatization and State Control in China* (pp. 111–124). London: Blackwell Publishing.

Liu, G.S. and Woo, W.T. (2001) How will ownership in China's industrial sector evolve with WTO accession? *China Economic Review,* 12, pp. 137–61.

Liu, Q. (2006) Corporate governance in China: Current practices, economic effects and institutional determinants. *CESifo Economic Studies,* 52(2), pp. 415–53.

Lockett, A. and Thompson, S. (2001) The resource-based view and economics. *Journal of Management,* 27, pp. 723–54.

Lu, T. (2002) Corporate governance in China. Working Paper: Institute of World Economics and Politics, Chinese Academy of Social Sciences.

Lu, T. (2005) Development of system of independent directors and the Chinese experience. Accessed at *http://www.cipe.org/China/development.htm.*

Lu, T. (2006a) Corporate governance in China. Accessed at *http://old.iwep.org .cn/cccg/pdf/*

Lu, T. (2006b) Corporate Governance Assessment Report of the Top 100 Chinese Listed Companies in 2006. Accessed at *http://en.iwep.org.cn/download/ upload_files/1uhuzc55kmcnio55gzfqen2e20070611163509.pdf.*

Lu, T., Zhong, J.Y. and Kong, J. (2009) How good is corporate governance in China? *China and World Economy,* 17(1), pp. 83–100.

Luo, Y. (2000) Dynamic capabilities in international expansion. *Journal of World Business,* 35(4), pp. 355–78.

Luo, Y. (2003) Industrial dynamics and managerial networking in an emerging market: The case of China. *Strategic Management Journal,* 24(13), pp. 1315–27.

Luo, Y. and Park, S.H. (2001) Guanxi and organizational dynamics: Organizational networking in Chinese firms. *Strategic Management Journal,* 22(5), pp. 455–77.

Luo, Y., Tan, J. and Shenkar, O. (1998) Strategic response to competitive pressures: The case of township and village enterprises in China. *Asia Pacific Journal of Management,* 15, pp. 33–50.

Lv, M.M. (14 December 2004) The introduction of development of Tiantong. *Zhejiang Electronics* newspaper.

Mace, M.L. (1971) Directors: myth and reality. Working Paper: Graduate School of Business Administration, Harvard University.

Mace, M.L. (1986) *Directors, Myth and Reality.* Boston, MA: Harvard Business School Press.

McDonald, K. (1993) Why privatization is not enough. *Harvard Business Review,* May–June, pp. 2–7.

Mahoney, J. (1995) The management of resource and the resource of management. *Journal of Business Research,* 33, pp. 91–102.

Mahoney, J. (2001) A resource-based theory of sustainable rents. *Journal of Management,* 27(6), pp. 651–60.

Mahoney, J. and Pandian, J.R. (1992) The resource-based view within the conversation of strategic management. *Strategic Management Journal,* 13(5), pp. 363–80.

Mako, W.P. and Zhang, C. (2003) *Management of China's State-Owned Enterprises Portfolio: Lessons from International Experience.* Beijing: World Bank Office.

Mallin, C. and Rong, X. (1998) The development of corporate governance in China. *Journal of Contemporary China,* 7(17), pp. 33–42.

March, J.G. (1991) Exploration and exploitation in organizational learning. *Organization Science,* 2, pp. 71–87.

Martinez, R.J. and Dacin, M.T. (1999) Efficiency motives and normative forces: Combining transactions costs and institutional logic. *Journal of Management,* 25(1), pp. 75–96.

Masini, A., Zollo, M. and Wassenhove, L. (2004) Understanding Exploration and exploitation in changing operating routines: The influence of industry and organizational traits. Working Paper: London Business School.

Mason, J. (2002) *Qualitative Researching* (2nd edn). London: Sage Publications.

Meyer, K. (2001) Institutions, transaction costs, and entry mode choice in Eastern Europe. *Journal of International Business Studies,* 32(2), pp. 357–67.

Meyer, K. (2007) Exploitation and exploration learning and the development of organizational capabilities: A cross-case analysis of the Russian oil industry. *Human Relations,* 60(10), pp. 1493–523.

Miao, W. (2007) *The Consumer Electronics Show in the Eyes of an Consumer,* from *www.lifeweek.com.cn.*

Miles, L. (2006) The application of Anglo–American corporate practices in societies influenced by Confucian values. *Business and Society Review,* 111(3), pp. 305–21.

Miles, R.E. and Snow, C.C. (1978) *Organizational Strategy, Structure, and Process.* New York: McGraw-Hill.

Miller, D. (1983) The correlates of entrepreneurship in three types of firms. *Management Science,* 29, pp. 770–91.

Miller, D. (1988) Relating Porter's business strategies to environment and structure: Analysis and performance implications. *Academy of Management Journal,* 31(2), pp. 280–308.

Miller, D. (1990) *The Icarus Paradox.* New York: Harper Collins.

Miller, D. and Friesen, P.H. (1986) Porter's generic strategies and performance. *Organization Studies,* 7, pp. 37–56.

Mina, G. and Perkins, F. (1997) *China's Transitional Economy – between Plan and Market.* Canberra: East Asia Analytical Unit.

Monks, R.A.G. (2001) Redesigning corporate governance structures and systems for the twenty-first century. *Corporate Governance – an International Review,* 9(3), pp. 142–7.

Monks, R.A.G. and Minnow, N. (2001) *Corporate Governance,* second edition. Oxford: Blackwell.

Morch, R., Yeung, B. and Zhao, M. (2005) China's lucky corporate governance. *Peking University Business Forum, September,* pp. 1–19.

Morgan, S.L. (1994) The impact of the growth of township enterprise on rural–urban transformation in China, 1978–1990. In A. Dutt, F. Costa, S. Aggarwal and A. Noble (Eds), *The Asian City: Processes of Development, Characteristics and Planning* (pp. 213–34). Boston, MA; London: Kluwer Academic Publishing.

Morgan, S.L. (2004) Professional associations and the diffusion of new management ideas in Shanghai, 1920–1930s: A research agenda. *Business and Economic History On-Line,* 2, pp. 1–24.

National Bureau of Statistics of China (2006), available at *http://www.stats.gov.cn.*

Naughton, B. (1995) *Growing out of the Plan: Chinese Economic Reform, 1978–1993.* New York: Cambridge University Press.

Naughton, B. (1997) *The China Circle: Economics and Electronics in the PRC, Taiwan, and Hong Kong*. Washington, DC: Brookings Institution Press.

Naughton, B. (2007) *The Chinese Economy: Transition and Growth*. London/Cambridge, MA: MIT Press.

Nee, V. (1996) The emergence of a market society: Changing mechanism of stratification in China. *American Journal of Sociology*, 101(4), pp. 908–49.

Nee, V. and Opper, S. (2006) China's politicized capitalism. Working Paper: Cornell University.

Nelson, R. (1990) Is there strategy in Brazil. *Business Horizons*, 33(4), pp. 15–23.

Newell, S. (1999) The transfer of management knowledge to China: Building learning communities rather than translating Western textbooks. *Education and Training*, 41(6/7), pp. 286–93.

Nicita, A. and Pagano, U. (2003) Corporate governance and institutional complementarities. Working Paper: University of Siena.

Nickell, S., Nicolitsas, D. and Dryden, N. (1997) What makes firms perform well. *European Economic Review*, 41, pp. 783–96.

North, D.C. (1990) *Institutions, Institutional Changes, and Economic Performance*. Cambridge; New York: Cambridge University Press.

OECD (5 April 2002) OECD/China, Industrial linkages: Trends and policy implications, *http://www.oecd.org/NewsArchives*.

OECD (2002) *China in the World Economy*. Paris: Organization For Economic Co-operation and Development.

OECD (2003) *OECD Investment Policy Reviews*. Paris: Organization For Economic Co-operation and Development.

OECD (2004) OECD Principles of Corporate Governance, *http://www.oecd.org/dataoecd/*.

Oliver, C. (1991) Strategic responses to institutional processes. *Academy of Management Review*, 16(1), pp. 145–79.

Oliver, C. (1997) Sustainable competitive advantage: Combining institutional and resource-based views. *Strategic Management Journal*, 18(9), pp. 697–713.

Operation newspaper, 11 April 2004.

Oswald, S. and Jahera, Jr. (1991) The influence of ownership on performance: An empirical study. *Strategic Management Journal*, 12, pp. 321–6.

Ozaki, H. (2004) Report of research on China. From *www.jcer.or.jp*.

Palepu, K., Khanna, T. and Vargas, I. (2005) Haier: taking a Chinese company global. ECCH: European Case Clearing House. Case Number: 9-706-4013.

Pan, S. and Park, A. (1998) Collective ownership and privatization of China's village enterprises. Working Paper: University of Michigan Business School.

Panda Electronics (2002) from *www.cnii.com.cn*.

Patton, A. and Baker, J. (1987) Why do not directors rock the boat? *Harvard Business Review* (65), pp. 10–12.

Pecht, M. and Chan, Y.C. (2004) *China's Electronics Industry*. Maryland: CALCE EPSC Press.

Peng, M. (1997) Firm growth in transitional economies: Three longitudinal cases from China, 1989–96. *Organization Studies*, 18(3), pp. 385–413.

Peng, M. (2002) Towards an institution-based view of business strategy. *Asia Pacific Journal of Management*, 19, pp. 251–67.

Peng, M., (2003) Institutional transitions and strategic choices. *Academy of Management Review*, 28(2), pp. 275–93.

Peng, M. (2004) Outside directors and firm performance during institutional transitions. *Strategic Management Journal*, 25(5), pp. 453–71.

Peng, M. (2005) Perspectives – from China strategy to global strategy. *Asia Pacific Journal of Management*, 22, pp. 123–41.

Peng, M. and Heath, P.S. (1996) The growth of the firm in planned economics in transition: Institutions, organizations, and strategic choice. *Academy of Management Review*, 21(2), pp. 492–528.

Peng, M. and Luo, Y. (2000) Managerial ties and firm performance in a transition economy: The nature of a micro–macro link. *Academy of Management Journal*, 43(3), pp. 486–501.

Peng, M., Tan, J. and Tong, T.W. (2004) Ownership types and strategic groups in an emerging economy. *Journal of Management Studies*, 41(7), pp. 1105–29.

Peng, M., Zhang, S. and Li, X. (2007) CEO duality and firm performance during China's institutional transitions. *Management and Organization Review*, 3(2), pp. 205–25.

Penrose, E.T. (1959) *The Theory of Growth of the Firm*. Oxford: Basil Blackwell.

Perotti, E.C., Sun, L. and Zou, L. (1999) State-owned versus township and village enterprises in China. *Comparative Economic Studies*, 41(2/3), pp. 151–79.

Peteraf, M.A. (1993) The cornerstone of competitive advantage: A resource-based view. *Strategic Management Journal*, 14, pp. 179–91.

Pfeffer, J. and Salanick, G.R. (1978) *The External Control of Organizations: A Resource Dependence Perspective*. New York: Harper and Row.

Pistor, K. (1999) Codetermination: A sociological model with governance externalities. In M. Blair and M. Roe (Eds.), *Employees and Corporate Governance* (pp. 163–193). Washington, DC: Brookings Institution Press.

Pollack, J.D. (1985) *The Chinese Electronics Industry in Transition*. Santa Monica: Rand.

Porter, M. (1980) *Competitive Strategy*. New York: Free Press.

Porter, M. (1985) *Competitive Advantage, Creating and Sustaining Superior Performance*. New York: Free Press.

Porter, M. (1986) Changing patterns of international competition. *California Management Review*, XXVIII(2), pp. 9–40.

PR Newswire. Hisense has achieved significant breakthrough for core technology through independent innovation (28 June 2005).

Prasad, E. (2004) China's growth and integration into the world economy: Prospects and challenges. Occasional Paper, International Monetary Fund, Washington, DC.

Priem, R.L. and Butler, J.E. (2001a) Is the resource-based 'view' a useful perspective for strategic management research? *Academy of Management Review*, 26(1), pp. 22–40.

Priem, R.L. and Butler, J.E. (2001b) Tautology in the resource-based view and the implications of externally determined resource value: Future comments. *Academy of Management View*, 26(1), pp. 57–66.

Prigge, S. (1998) A survey of German corporate governance. In K.J. Hopt, H. Kanda, M.J. Roe, E. Wymeersch and S. Prigge (Eds.), *Comparative*

Corporate Governance: the State of the Art and Emerging Research (pp. 943–1045). Oxford: Clarendon Press.

Pye, L. (1992) Social science theories in search of Chinese realities. *China Quarterly,* 132, pp. 1161–70.

Qian, Y. (2002) How reform worked in China. Working Paper: Department of Economics, University of California, Berkeley.

Qu, L.Z. (2004) The rapid growth of electronics industry in China in 2003. Zhongguo Xin Jingji Touzi yu Gao Keji *(New Business Investment and Hi-Technology in China),* 3, pp. 26–8.

Qu, Q. (2003) Corporate governance and state-owned shares in China: Listed companies. *Journal of Asian Economics,* 14(5), pp. 771–83.

Rajagopalan, N. and Zhang, Y. (2008) Corporate governance reforms in China and India: Challenges and opportunities. *Business Horizons,* 51(1), pp. 55–64.

Rawski, T.G. (2007) Social capabilities and Chinese economic growth. In W. Tang and B. Holzner (Ed.), *Social Change in Contemporary China: C.K. Yang and the Concept of Institutional Diffusion* (pp. 89–103). Pittsburgh PA, University of Pittsburgh Press.

Rindfleisch, A. and Heide, J.B. (1997) Transaction cost analysis: Past, present, and future application. *Journal of Marketing,* 61(4), pp. 30–54.

Roe, M.J. (1990) Political and legal restraints on ownership and control of public companies. *Journal of Financial Economics,* 27, pp. 7–41.

Roe, M.J. (1997) The political roots of American corporate finance. *Journal of Applied Corporate Finance,* 9(4), pp. 8–21.

Salmon, W.J. (2000) *Empowering the Board. Harvard Business Review on Corporate Governance.* Harvard Business School Press.

Schlevogt, K.-A. (2000) Doing business in China Part I: The business environment in China – getting to know the next century's superpower. *Thunderbird International Business Review,* 42(1), pp. 85–111.

Schneider-Lenne, E.R. (1992) Corporate governance in Germany. *Oxford Economic Policy,* 8(3), pp. 11–23.

Scott, W.R. (1995) *Institutions and Organizations.* Thousand Oaks, CA: Sage Publications.

Scott, W.R. (2002) The changing world of Chinese enterprise: An institutional perspective. In A.S. Tsui and C.M. Lau (Ed.), *The Management of Enterprises in the People's Republic of China* (pp. 59–78). Boston: Kluwer Academic Publisher.

Shen, S.B. and Jia, Ding. (2004) Will the independent director institution work in China? *Loyola Los Angeles International and Comparative Law Review,* 27, pp. 223–48.

Shepherd, W. (1988) Public enterprises: Criteria and cases. In H.W. d. Jong (Ed.), *The Structure of European Industry* (pp. 355–388). Boston, MA: Kluwer Academic Publisher.

Shi, Q. and Zhao, J. (2001) *The Report of Industrial Development in China, 2000.* Beijing: Zhongguo qing gong ye chu ban she.

Shi, Y. (1998) *Chinese Firms and Technology in the Reform Era.* London and New York: Routledge.

Shleifer, A. and Vishny, R. (1994) Politicians and firms. *Quarterly Journal of Economics, November,* pp. 995–1025.

Shleifer, A. and Vishny, R.W. (1997) A survey of corporate governance. *Journal of Finance*, VII(2), pp. 737–83.

Sicular, T. (1995) Why state-owned enterprises choose to make loss. *Economic Research Journal*, 5, pp. 21–8.

Simon, D.F. (1992) Sparking the electronics industry. *China Business Review*, 19, pp. 22–8.

Simon, D.F. (1996) From cold to hot: China struggles to protect and develop a world class electronics industry. *China Business Review*, 23, pp. 8–16.

Singh, A. (2002) Competition and competition policy in emerging markets: International and developmental dimensions. Working Paper: Queen's College, University of Cambridge.

Sit, V.F.S. (1983) Collective industrial enterprises in the People's Republic of China. *Human Geography*, 65(2), pp. 85–94.

Smith, K., Guthrie, J. and Chen, M.J. (1989) Strategy, size and performance. *Organization Studies*, 10, pp. 63–81.

Song, L.G. and Yao, Yang. (2003) Impacts of privatization on firm performance in China. Working Paper, China Centre for Economic Research, Peking University.

Song, F., Yuan, P. and Gao, F. (2006) Does a large state shareholder affect the governance of Chinese Board of Directors? Working paper, Tsinghua University.

Stainer, A. and Heap, J. (1996) Attachment H: Growing importance of non-financial performance measures. *Management Services*, 40(7), pp. 10–12.

Statistical Bureau of Zhejiang Province (2004) Survey on transferred enterprises in Zhejiang Province. Report in *Qiangjiang Evening Post (Qianjiang wanbao)*, 29 September.

Su, J. and Jefferson, G. (2003) A theory of decentralized privatization: Evidence from China. Working Paper: Graduate School of International Economics and Finance, Brandeis University.

Sun, L. (2002a) Anticipatory ownership reform driven by competition: China's Township – Village and Private Enterprises in the 1990s. *Comparative Economic Studies*, XLII, 3 (Fall), pp. 49–75.

Sun, L. (2002b) Fading out of local government ownership: Recent ownership reform in China's township and village enterprises. *Economic Systems*, 26, pp. 249–69.

Sun, Q., Tong, W.H.S. and Tong, J. (2002) How does government ownership affect performance? Evidence from China's privatization experience. *Journal of Business Finance and Accounting*, 29((1) and (2)), pp. 1–27.

Tam, O.K. (Ed.) (1995) Financial reform in China. London; New York: Routledge.

Tam, O.K. (1999) *The Development of Corporate Governance in China*. Cheltenham: Edward Elgar Publishing Limited.

Tam, O.K. (2000) *Corporate Governance and the Future of China's Corporate Development*. Paper presented at the Conference paper for EAMSA 2000 Conference, Singapore.

Tam, O.K. (2002) Ethical issues in the evolution of corporate governance in China. *Journal of Business Ethics*, 37, pp. 303–20.

Tan, J. (1996) Regulatory environment and strategic orientations in a transitional economy: A study of Chinese private enterprise. *Entrepreneurship: Theory and Practice*, 21(1), pp. 31–47.

Tan, J. (2002) Impact of ownership type on environment strategy linkage and performance: Evidence from a transitional economy. *Journal of Management Studies*, 39(3), pp. 333–54.

Tan, J. (2005) Venturing in turbulent water: A historical perspective of economic reform and entrepreneurial Transformation. *Journal of Business Venturing*, 20(5), pp. 689–704.

Tan, J. and Litschert, R.J. (1994) Environment–strategy relationship and its performance implications: An empirical study of the Chinese electronics industry. *Strategic Management Journal*, 15(1), pp. 1–20.

Tan, J. and Tan, D. (2003) A dynamic view of organizational transformation: The changing face of Chinese SOEs under transition. *Journal of Leadership and Organizational Studies*, 10(2), pp. 98–112.

Tan, J. and Tan, D. (2004) Environment Strategy co-evolution and co-alignment: A strategic model of Chinese SOEs under transition. *Strategic Management Journal*, 26(2), pp. 141–57.

Tang, J. and Ma, L.J.C. (1985) Evolution of urban collective enterprises in China. *China Quarterly*, 104, pp. 614–40.

Tang, J. and Ward, A. (2003) *The Changing Face of Chinese Management*. London; New York: Routledge.

Tang, X.Q. (2002) The international expansion of Chunlan. Nanjing Shehui Kexue (*Nanjing Social Science*), 11, pp. 91–3.

Teece, D.J., Pisano, G. and Shuen, A. (1997) Dynamic capabilities and strategic management. *Strategic Management Journal*, 18, pp. 509–33.

Tenev, S., Zhang, C. and Brefort, L. (2002) *Corporate Governance and Enterprise Reform in China*. Washington, DC: World Bank and the International Finance Corporation.

Thomsen, S. and Pedersen, T. (2000) Ownership structure and economic performance in the largest European companies. *Strategic Management Journal*, 21(6), pp. 689–705.

Thornhill, S. and White, R.E. (2007) Strategic purity: A multi-industry evaluation of pure vs hybrid business strategy. *Strategic Management Journal*, 28, pp. 553–61.

Tian, J. and Lau, C.M. (2001) Board composition, leadership structure and performance in Chinese shareholding companies. *Asia Pacific Journal of Management*, 18, pp. 245–63.

Tong, Ying. (27 May 2004) The status quo of independent directors in China (*Zhongguo Duli Dongshi Xianzhuang*). Shanghai Zhengquan Bao (*Shanghai Securities* newspaper).

Tsui, A. (2004) Contributing to global management knowledge: A case for high quality indigenous research. *Asia Pacific Journal of Management*, 21, pp. 491–513.

Tsui, A., Schoonhoven, C., Meyer, M., Meyer, C., Lau, C. and Milkovich, G. (2004) Organization and management in the midst of societal transformation: People's Republic of China. *Organization Science*, 15(2), pp. 133–44.

Tüngler, G. (2000) The Anglo–American board of directors and the German supervisory board – Marionettes in a puppet theatre of corporate governance or efficient controlling devices? *Bond Law Review*, 12 (2), pp. 230–6.

Turnbull, S. (1994) Competitiveness and corporate governance. *Corporate Governance: An International Review*, 2(2), pp. 90–6.

Turnbull, S. (1997) Corporate governance: Its scope, concerns and theories. *Corporate Governance: An International Review*, 5(4), pp. 180–205.

Turnbull, S. (2000) Why unitary boards are not best practice: A case for compound boards. Working Paper: Macquarie University Graduate School of Management, Sydney.

Uhlenbruck, K., Meyer, K.E. and Hitt, M.A. (2003) Organizational transformation in transition economies: Resource-based and organizational learning perspective. *Journal of Management Studies*, 40(2), pp. 257–82.

Urban, G.L. and Star, S. (1991) *Advanced Marketing Strategy*. NJ: Prentice Hall.

Vashist, D. (2004) Chinese state-owned enterprises: The challenges. Working Paper: ICFAI Business School Case Development Centre, India.

Vickers, J. and Yarrow, G. (1991) Economic perspectives on privatization. *Journal of Economic Perspective*, 5(2), pp. 111–32.

Walder, A. (1994) Evolving property rights and their political consequences. In V. Milor (Ed.), *Changing Political Economies: Privatization in Post-Communist and Reforming Communist States*. Boulder, CO; London: Lynne Rienner, pp. 53–66.

Walder, A. (1995a) 'The quiet revolution from within: Economic reform as a source of political decline.' In A.G. Walder (ed.), *The Waning of the Communist State: Economic Origins of Political Decline in China and Hungary*, Berkeley: University of California Press, pp. 1–24.

Walder, A. (1995b) Local governments as industrial firms: An organizational analysis of China's transitional economy. *American Journal of Sociology*, 101(2), pp. 263–301.

Walder, A. (2003) Elite opportunity in transitional economies. *American Sociological Review*, 68(6), pp. 899–916.

Wang, H. (1994) *The Gradual Revolution*. New Brunswick, NJ: Transaction Publishers.

Wang, Y. and Yao, Y. (2001) *Market Reforms, Technological Capabilities, and the Performance of Small Enterprises in China*: International Bank for Reconstruction and Development/World Bank.

Wang, Z.K. (2006) The growth of China's private sector: A case study of Zhejiang Province. *China and World Economy*, 14(3), pp. 109–20.

Warner, M. (2002) The future of Chinese management. In M. Warner (Ed.), *The Future of Chinese Management* (pp. 205–23). London: Frank Cass.

Wei, G. (2002) The transformation of Chunlan. *Knowledge Economy*, 8, pp. 11–6.

Wei, Y. (2003) An overview of corporate governance in China. *Syracuse Journal of International Law and Commerce*, 30(1), pp. 23–48.

Wei, G. and Geng, M. (2008) Ownership structure and corporate governance in China: Some current issues. *Managerial Finance*, 34(2), pp. 934–52.

Weick, K.E. and Westley, F. (1996) Organizational learning: Affirming an oxymoron. In S.R. Clegg, C. Hardy and W.R. Nord (Eds.), *Handbook of Organization Studies* (pp. 440–58). London: Sage.

Weinstein, M. (2008) The independent director requirement and its effects on the foreign investment climate in China: Progress or regress? *Business Law Brief*, Spring, pp. 35–9.

Wen, M. (2002) Competition, ownership diversification and industrial growth in China. Working Paper: Australian National University.

Wernerfelt, B. (1984) A resource-based view of the firm. *Strategic Management Journal*, 5(2), pp. 171–80.

White, S. (2000) Competition, capabilities, and the make, buy, or ally decision of Chinese state-owned firms. *Academy of Management Journal*, 43(3), pp. 324–41.

White, S. and Xie, W. (2006) Lenovo's pursuit of dynamic strategic fit. in A.S. Tsui, Y. Bian and L. Cheng (Eds.), *China's Domestic Private Firms* (Vol. 1, pp. 277–96). Armond, NY: M.E. Sharpe.

Whitley, R. and Czaban, L. (1998) Institutional transformation and enterprise change in an emergent capitalist economy: The case of Hungary. *Organization Studies*, 19(2), pp. 259–80.

Williams, J.R. (1992) How sustainable is your competitive advantage? *California Management Review*, 34(3), pp. 29–51.

Williamson, O.E. (1975) *Markets and Hierarchies: Analysis and Antitrust Implications*. New York: Free Press.

Williamson, O.E. (1984) Corporate governance. *Yale Law Journal*, 93, pp. 1197–230.

Williamson, O.E. (1985) *The Economic Institutions of Capitalism: Firms, Markets, Relational Contracting*. London: Collier Macmillan.

Williamson, O.E. (1994) Strategizing, economizing, and economic organization. In R.P. Rumelt, D.E. Schendel and D.J. Teece (Eds.), *Fundamental Issues in Strategy* (pp. 361–402). Boston, MA: Harvard Business School Press.

Wiseman, R., Dykes, B., Weidlich, R. and Franco-Santos, M. (2006) Pursuing a dual strategy of exploitation and exploration in Central and Eastern Europe. Working Paper: The Eli Broad Graduate School of Management, Michigan State University.

Wittington, R. and Mayer, M. (2000) *The European Corporations: Strategy Structure, and Social Science*. New York: Oxford University Press.

Wong, S.L. (1985) The Chinese family firm: A model. *British Journal of Sociology*, 36(1), pp. 58–72.

World Bank (2000) *China's Emerging Private Enterprises*, Washington, D.C.

World Bank Group. (2001) *Reform of China's State-owned Enterprises, http://worldbank.org/html/prddr/trans*.

Wortzel, H. and Wortzel, L. (1989) Privatization: Not the only answer. *World Development*, 17(5), pp. 633–41.

Wright, M., Filatotchev, I., Hoskisson, R.E. and Peng, M.W. (2005) Strategy research in emerging economies: challenging the conventional wisdom. *Journal of Management Studies*, 42(1), pp. 1–33.

Wu, C. and Li, D.D. (2006) Firm behaviour in a mixed market. In A. Tsui, Y. Bian and L. Cheng (Eds.), *China's Domestic Private Firms* (pp. 171–95): Armonk, NY: M.E. Sharpe.

Wu, Y.J. and Tang, J.Y. (1999) The analysis of competitive advantage of Hisense. *Zhongguo Gongye Jingji* (*China Industrial Economy*), 5, pp. 73–7.

Wymeersch, E. (1998) A status report on corporate governance rules and practices in some continental European States. In Hopt K.J., Kanda H., Roe M.J., Wymeersch E. and Prigge S. (Ed.), *Comparative Corporate Governance: the State of the Art and Emerging Research* (pp. 1045–1200). Oxford: Clarendon Press.

Xiao, K. and Pei, F. (2003) The transformation of Silan. *China Science and Technology Zone*, 9, pp. 26–7.

Xiao, Z.D., Dahya, Jay. and Lin, Z.J. (2004) A grounded theory exposition of the role of the supervisory board in China. *British Journal of Management*, 15, pp. 39–55.

Xie, W. and Wu, G.S. (2003) Differences between learning processes in small tigers and large dragons: Learning processes of two color TV (CTV) firms within China. *Research Policy*, 32(8), pp. 1463–79.

Xu, X. and Wang, Y. (1997) *Ownership Structure, Corporate Governance, and Firm's Performance: The Case Of Chinese Stock Companies*: Amherst College and The World Bank.

Xu, X. and Wang, Y. (1999) Ownership structure and corporate governance in Chinese stock companies. *China Economic Review*, 10, pp. 75–98.

Yang, R. and Zhang, Y. (2003) Globalization and China's SOEs Reform. Conference Paper for 'Sharing the Prosperity of Globalization', Helsinki, Finland. 6–7 September 2003.

Yin, R. (1984) *Case Study Research, Design and Methods*. London: Sage Publications.

Yin, R. (2003) *Applications of Case Study Research*. London; New Delhi: Sage Publications.

Yin, X. (1998) The macroeconomic effects of waiting workers in the Chinese economy. *Journal of Comparative Economics*, 26(1), pp. 156–64.

Young, S. (1989) Policy, practice and the private sector in China. *Australian Journal of Chinese Affairs*, 21, pp. 57–80.

Yu, K. (2005) Nanjing Government helps Panda out of the plight. *Capital Market*, 4, pp. 42–3.

Yuan, J. (2007) Formal convergence or substantial divergence? Evidence from adoption of the independent director system in China. *Asian-Pacific Law and Policy Journal*, 9(1), pp. 72–104.

Zhang, Y. and Parker, D. (2001) *The Impact of Ownership on Management and Structures in the Chinese Electronics Industry: A Questionnaire Based Study*. Discussion Paper, Aston University, Birmingham.

Zhang, Y.-F. and Parker, D. (2004) Labour and total factor productivity in the Chinese electronics industry in the 1990s. *International Review of Applied Economics*, 18(1), pp. 79–1.

Zhou, K.Z. and Li, C.B. (2007) How does strategic orientation matter in Chinese firms? *Asia Pacific Journal of Management*, 24, pp. 447–66.

Zhou, X. (2001) Rethinking property rights as a relational concept: Exploration in China's transitional economy. Working Paper: Department of Sociology, Duke University.

Zollo, M. and Winter, S. (2001) Deliberate learning and the evolution of dynamic capabilities. *Organization Science*, 13(3), pp. 339–51.

III Useful links

1. *http://www.news.xinhuanet.com*: Xinhua News website.
2. *http://www.xinhua.com*: Xinhua News Agency website.
3. *http://www.chinaretailonline.com*: China Retail website.
4. *http://www.csrc.gov.cn*: China Securities Regulatory Commission website.
5. *http://www.adbi.org*: Asian Development Bank Institute website.
6. *http://www.hisense.com*: Hisense Company website.
7. *http://www.panda.cn*: Panda Company website.
8. *http://www.hdeg.com*: Huadong Company website.
9. *http://www.haier.com*: Haier Company website.
10. *http://www.chunlan.com*: Chunlan Company website.
11. *http://www.yahnkon.com*: Yankon Company website.
12. *http://www.tiantong.com*: Tiantong Company website.
13. *http://www.sina.com*: SINA website.
14. *http://www.fdi.gov.cn*: Investment in China website.
15. *http://www.most.gov.cn*: Ministry of Science and Technology of the People's Republic of China website.
16. *http://www.sse.com*: Shanghai Stock Exchange website.
17. *http://www.szse.cn*: Shenzhen Stock Exchange website.
18. *http://www.hkex.com.hk*: Hong Kong Stock Exchange website.
19. *http://www.oecd.org*: OECD website.
20. *http://www.cnii.com.cn*: China Information Industry website.
21. *http://www.zjlca.com*: Zhejiang List Company website.
22. *http://www.worldbank.org*: World Bank website.

IV List of interviewees from companies and non-company departments

This section lists the interviewees who were the informants for the case studies. The interviewees are organized alphabetically for each company and non-company.

Part 1. Interviewees from each company

Company	Sequence number	Gender	Interviewee role	Years with Co.	Years on current position	Date of interview	Place of interview
Hisense	1	M	Non-executive director	15	8	07/2007	Qingdao, Shandong
	2	M	Senior manager	12	11	05/2006	Qingdao, Shandong
	3	M	Non-executive director	23	8	07/2007	Qingdao, Shandong
	4	F	Supervisory board member	11	5	05/2006	Qingdao, Shandong
	5	M	Senior manager	18	8	04/2006	Jinan, Shandong
	6	M	Senior manager	12	11	05/2006	Qingdao, Shandong
	7	M	Independent director	N/A	6	04/2006	Jinan, Shandong
Panda	8	M	Senior manager	18	10	06/2006	Nanjing, Jiangsu
	9	M	Senior manager	25	5	06/2006	Nanjing, Jiangsu
	10	F	Non-executive director	9	9	05/2007	Jinan, Shandong
	11	F	Supervisory board member	28	4	05/2007	Jinan, Shandong
	12	M	Senior manager	16	10	06/2006	Nanjing, Jiangsu
	13	M	Independent director	N/A	7	04/2007	Nanjing, Jiangsu
	14	M	Non-executive director	27	13	04/2007	Nanjing, Jiangsu
Huadong	15	M	Supervisory board member	30	5	04/2007	Nanjing, Jiangsu
	16	F	Senior manager	30	4	06/2006	Nanjing, Jiangsu
	17	M	Non-executive director	25	4	05/2007	Jinan, Shandong
	18	F	Senior manager	15	5	06/2006	Nanjing, Jiangsu

19	M	Senior manager	23	5	04/2007	Nanjing, Jiangsu
20	M	Board secretary Executive director	18	4	06/2006	Nanjing, Jiangsu
21	M	Independent director	N/A	3	05/2007	Jinan, Shandong
Haier 22	M	Executive director	12	4	06/2007	Jinan, Shandong
23	M	Supervisory board member	N/A	4	05/2006	Qingdao, Shandong
24	M	Board secretary	9	4	05/2006	Qingdao, Shandong
25	M	Executive director	22	4	07/2007	Qingdao, Shandong
26	M	Senior manager	12	6	05/2006	Qingdao, Shandong
Chunlan 27	F	Non-executive director	28	5	05/2007	Jinan, Shandong
28	M	Senior manager	19	5	05/2007	Jinan, Shandong
29	M	Board secretary	20	5	06/2006	Taizhou, Jiangsu
30	M	Senior manager and executive director	19	4	06/2006	Taizhou, Jiangsu
31	M	Supervisory board member	17	5	04/2007	Taizhou, Jiangsu
32	M	Independent director	N/A	4	04/2007	Taizhou, Jiangsu
Yankon 33	F	Senior manager	12	5	06/2007	Shangyu, Zhejiang
34	M	Supervisory board member	10	5	06/2007	Shangyu, Zhejiang
35	M	Senior manager Executive director	10	5	05/2007	Jinan, Shandong
36	M	Senior manager, Executive director	10	5	07/2006	Shangyu, Zhejiang
37	F	Board secretary Executive director	6	3	07/2007	Shangyu, Zhejiang
Silan 38	F	Board secretary	12	3	07/2006	Hangzhou, Zhejiang
39	M	Executive director	11	11	06/2007	Hangzhou, Zhejiang
40	M	Executive director and Deputy general manager	8	5	06/2007	Hangzhou, Zhejiang

Part 1. Interviewees from each company (Cont'd)

				11	11		
	41	M	Executive director	11	11	06/2007	Hangzhou, Zhejiang
	42	M	Senior manager	8	6	07/2006	Hangzhou, Zhejiang
	43	M	Supervisory board member	8	5	06/2007	Hangzhou, Zhejiang
Tiantong	44	M	Executive director	6	3	07/2006	Haining, Zhejiang
	45	M	Executive director	8	3	06/2007	Haining, Zhejiang
	46	F	Board secretary	3	3	07/2006	Haining, Zhejiang
	47	M	Senior manager	3	3	07/2007	Jinan, Shandong
	48	M	Independent director	N/A	6	06/2007	Hangzhou, Zhejiang
	49	M	Senior manager	12	3	06/2007	Haining, Zhejiang
	50	M	Supervisory board member	7	7	05/2007	Jinan, Shandong

Part 2. Interviewees from non-company departments

Department	Sequence number	Gender	Interviewee role	Years with company	Years in current position	Date of interview	Place of interview
Policy and Regulation Office, Shandong SASAC	51	M	Deputy director	15	3	07/2006	Jinan, Shandong
Shandong Securities Co.	52	M	Senior consultant	14	6	07/2006	Jinan, Shandong
Qingdao Pale Consulting Co.	53	M	Senior consultant	20	8	05/2006	Qingdao, Shandong
Shandong Electronics Bureau	54	M	Vice president	34	10	07/2006	Jinan, Shandong
Huarong Asset Management Co. Beijing	55	M	Assistant to general manager	18	5	04/2006	Beijing
Shandong Securities Co.	56	M	Senior manager	12	5	04/2006	Jinan, Shandong

Index

Printed and bound by CPI Group (UK) Ltd, Croydon, CR0 4YY

08/05/2025

01864972-0001